AF323579

QUALITY UP, COSTS DOWN

A Manager's Guide to Taguchi Methods and QFD

QUALITY UP, COSTS DOWN

A Manager's Guide to Taguchi Methods and QFD

Edited by

William E. Eureka

Nancy E. Ryan

ASI Press
Tools for the Competitive Edge

IRWIN
Professional Publishing
Burr Ridge, Illinois
New York, New York

© AMERICAN SUPPLIER INSTITUTE, 1995

Project editor: Karen J. Nelson
Production manager: Laurie Kersch
Interior designer: Michael Warrell
Cover designer: Tim Kaage
Compositor: Precision Typographers
Typeface: 12/15 Bookman
Printer: Book Press

Library of Congress Cataloging-in-Publication Data

Quality up, costs down : a manager's guide to Taguchi methods and QFD
/ edited by William E. Eureka, Nancy E. Ryan.
    p.   cm.
   Rev. ed. of: Taguchi methods and QFD. c1988.
   Includes bibliographical references (p.   ) and index.
   ISBN 0-7863-0218-6
   1. Taguchi methods (Quality control) 2. Quality function
deployment. I. Eureka, William E. II. Ryan, Nancy E.
III. Taguchi methods and QFD.
  TS156.T344 1995
  658.5'62—dc20   94–14951

Printed in the United States of America
1 2 3 4 5 6 7 8 9 0 BP 1 0 9 8 7 6 5 4

Preface

"GM plans to embrace 'value pricing' in the 1994 model year, in the hope that offering better-equipped cars at lower prices will raise GM's market share and cut manufacturing costs."
—"What's News," *The Wall Street Journal*, July 12, 1993

This quote is about increased quality and reduced costs—and GM's not the only company doing it.

"Americans innovate; Japanese manufacture. For decades, that reality has undercut US companies as Japanese firms took over one field after another, from TVs to memory chips. Now a new generation of US technology is about to hit the market, and Japan's firms are again gearing up to make the devices."
—"US, Japan Focusing on Electronic Gear,"
The Wall Street Journal, July 12, 1993

This quote concerns time to market—and it's never been more critical.

"Interest in Taguchi Methods[TM] *and Quality Function Deployment (QFD) is snowballing as American companies that have implemented them share their experiences and success stories through case studies and public presentations. There is already ample documentation showing how the application of these methodologies has enabled American companies to improve product quality and reduce costs and product-development times while gaining market share in an increasingly competitive global marketplace."*
—John P. Kennedy, director of publishing, ASI Press
Taguchi Methods and QFD:
How's and Why's for Management

As indicated in this quote, Taguchi Methods and QFD have never been more critically needed.

With the above thoughts in mind, the second edition of *Taguchi Methods and QFD: How's and Why's for Management,* first published by the ASI Press in 1988, has been aptly retitled *Quality Up, Costs Down: A Manager's Guide to Taguchi Methods and QFD.* The search for papers and essays to fill this second edition gave substance to Kennedy's words. The use of Taguchi Methods—combined engineering and statistical methods that achieve rapid improvements in cost and quality by optimizing product design and manufacturing processes—and QFD—a system for translating customer requirements into appropriate company requirements at each stage of the product development cycle, from research and development to engineering, manufacturing, marketing, sales, and distribution—has indeed snowballed in the United States.

The selections that follow, many of which were prepared for and presented at recent symposia and conferences on Taguchi Methods and QFD prior to publication here, provide valuable insight into the need for and implementation of these methodologies in America. Three of the selections also appeared in the first edition but are reprinted here as well due to their timely message and timeless quality. In addition, two of the selections are excerpts from the second editions of two books first published by the ASI Press in the late 1980s, *Quality by Design: Taguchi Methods and U.S. Industry* and *The Customer-Driven Company: Managerial Perspectives on QFD,* both recently reissued by ASI and Irwin Professional Publishing, Burr Ridge, Homewood, Illinois.

Quality by Design examines Dr Genichi Taguchi's methodology and philosophy from both a historical and a technical perspective, as well as its increasing influence in the United States. *The Customer-Driven Company,* an introduction to QFD and its related benefits, includes the insights and experiences of senior managers from American companies that have applied the methodology.

I sincerely wish to thank the authors of all the selections—those of papers and essays from the first edition as well as from the second—for granting us permission to reprint their work here. Both Taguchi Methods and QFD improve as they are shared, with the insight and experience gleaned from one company cumulatively contributing to the work of the next.

William E. Eureka

Acknowledgments

The ASI Press and Irwin Professional Publishing are proud to have the opportunity to present these important papers and wish to acknowledge the following authors for permitting their publication:

Diane M. Byrne, principal, ITEQ International, Livonia, Michigan.

Dr Don Clausing, Bernard M. Gordon Adjunct Professor of Manufacturing Innovation and Practice, Massachusetts Institute of Technology, Cambridge, Massachusetts.

Robert J. Dika, manager, cross platform quality planning, corporate customer satisfaction vehicle quality office, Chrysler Corp., Auburn Hills, Michigan.

Lance A. Ealey, consultant, McKinsey and Company, Cleveland, Ohio.

William E. Eureka, learning coach, Herman Miller Inc., Zeeland, Michigan.

Kurt R. Hofmeister, director, American Supplier Institute, Dearborn, Michigan.

Robert Moesta, principal, Pedi, Moesta & Associates, Grosse Pointe Farms, Michigan.

Jim Quinlan, principal, ITEQ International, Livonia, Michigan.

Nancy E. Ryan, free-lance writer, Oakland, Michigan.

Robert H. Schaefer, director of systems engineering methods, General Motors Corp., Warren, Michigan.

Shin Taguchi, vice president, American Supplier Instutite, Allen Park, Michigan.

David P. Williams, president and chief operating officer, The Budd Co., Troy, Michigan.

Contents

Chapter One

Taguchi: A Managerial Perspective

David P. Williams

Doing business in today's marketplace is challenging, but it's also very exciting. We know that success today and in the future requires recognition, understanding, and utilization of every improvement strategy that fits our individual needs.

The pursuit of world-class quality seems to require movement through several phases. Phase one might be recognizing the need for change. Phases two through three, four, and maybe even five could be thought of as progressing through management commitment, involvement, communication, and all those things that deal with the desire to seek continuous quality improvement. The desire for world-class quality and the reaching of that destination are miles apart. Now we're looking for the tools that will pave the road.

One such tool is Dr Genichi Taguchi's Quality Engineering methodology. Taguchi Methods™ reduce costs

---

David P. Williams is president and chief operating officer of The Budd Co., Troy, Michigan.

through reduction of variation, which will always result in improved quality, and allow designs and processes to be optimized in a way that makes them insensitive to factors beyond the manufacturer's control. Now that's about as technical as I get. All I know is that it has helped us improve quality and, at the same time, reduce costs associated with scrap, inspection, and rework. Moreover, it has delivered quick results with little investment.

This past summer, The Budd Co. launched its third companywide Taguchi Methods design of experiments seminar. About 40 engineers attended five days of off-site classroom training. At that time, we completed training approximately 120 people from our divisions and plants. The two previous seminars, held in October 1986 and March 1987, resulted in more than 20 major projects dealing with sheet-metal stamping, spot welding of galvanized steel, spinning of wheel disks, grinding of brake rotors, and molding and assembly bonding of plastic parts.

We have a simple formula for success with our Taguchi Methods training that I'd like to share: Our seminars are 5 + 1 + 2. We schedule a five-day off-site seminar conducted by the American Supplier Institute (ASI), Inc. The attendees are divided into product teams. These teams are subjected to classroom lecture, but they also initiate an experiment during the session. The team members continue this experiment at their respective divisions. Two months later, we have ASI consult on the experiment for one day at each of our divisions. In another two months, the teams return to an off-site location for two days, prepared to present the results of their experiment for critique.

We started with a small number of enthusiasts who attempted to carry the banner and promote the concept.

I'll have to admit that there was a fair amount of skepticism among management and early students alike. Successful completion of case studies and application to actual product problems, however, has generated overwhelmingly positive attitudes.

Once we moved the concept out of the classroom and put it into practice, a creative and competitive interest took over that has resulted in at least 20 major project successes in less than a two-year period, as the following examples illustrate.

Our stamping and frame division recently completed a study that investigated the relationship between press variables and stamping quality for the first draw-die operations on a door inner panel. Plans call for more experiments in the near future, with the ultimate goal being the routine use of Taguchi Methods during die tryout in order to provide quick determination of the optimum settings. This should improve our die transition times, as well as the consistency of our first-piece quality.

Another study within that division addressed the classic challenge of indirect spot welding of sheet-metal hem flanges. The challenge to which I'm referring is the maintenance of weld integrity while satisfying surface-appearance requirements. This was an interesting experiment for our people: You can assure either weld quality or surface quality at the expense of the other. Optimization of the process to satisfy both requirements was an experiment particularly well suited for application of Taguchi Methods.

The end results were favorable in that we were able to reduce our internal repair and scrap costs significantly. It's also important to recognize, however, the improvement process involving this particular experiment. The experiment was initiated by a team of young engineers who were mechanically oriented but certainly not seasoned or even trained engineers. With five days of Taguchi Methods training, they went back to their plant to attempt to tell people in a company with 75 years of welding background how to improve. When the results proved favorable, the plant experts and skeptics were sure that their infinite job knowledge could have resolved the problem if someone had listened. Yes, job knowledge is important. In this case, however, Taguchi Methods provided the facts and discipline necessary to make sound judgments related to process changes. Moreover, the data were collected by hourly personnel and the procedure was explained so that there wasn't idle involvement, but involvement with a degree of understanding.

Our wheel and brake division reports success in applying Taguchi Methods to the manufacture of heavy truck wheels and the grinding and assembly of brake rotors. One study focusing on the assembly module for a large unicast rotor was designed to optimize eight control variables concerning five inspection characteristics. The results established control settings that reduced rework and scrap by 87.5 percent. Yet another study was aimed at determining process settings for a computer numerically-controlled spinning machine that had been taken off-line because of poor performance. The experiment resulted in the machine being placed back into production. That same machine is currently spinning wheel disks with a high degree of consistency.

Our plastics division has experienced similarly favorable results applying Taguchi Methods to the molding and assembly of automotive body panels, diesel-engine components, and water skis made of fiberglass-reinforced plastics. A recent study was aimed at minimizing the scrap rate of a molding process used to make an engine-valve cover. The experiment considered 14 process-control settings and four-part characteristics that had been the cause of rejects in the past. The results identified control settings that may eliminate the incidence of one undesirable characteristic altogether and significantly reduce the recurrence of the others.

In addition to fine-tuning production processes so that they produce parts with less variability, Taguchi Methods provide other important benefits to The Budd Co.:

- They enhance communication between functional groups by providing a uniform communication format.
- They help the company determine the best methods of process control by rapidly determining the optimum setting for a number of control variables.
- They permit a more efficient method for changing processes, allowing fewer adjustments and more predictable effects from each adjustment.
- They reduce process costs with essentially no capital expenditures.
- They can be applied to a variety of processes.

When dealing with similar circumstances in the past, our people—while knowledgeable and dedicated—would tend to react to their gut feelings. Everyone would want to jump in and tear the machine or equipment apart without significant analysis of the problem. Taguchi Methods have allowed us to achieve a better understanding of process needs and requirements.

In the past, we wrote procedures that were seldom read and almost never followed. Taguchi Methods have taught us to establish process parameters that are now set and controlled.

Analysis-of-variance tables and Signal-to-Noise Ratios are terms familiar to practitioners of Taguchi Methods, but frankly, they're foreign to me. Competition and the need for continuous planned improvement, however, aren't foreign to me or The Budd Co.

We see Taguchi Methods as a tool to improve the performance of a process as it's used in day-to-day production. I applaud ASI for its leadership in providing the technical expertise for—as well as a nontechnical understanding of—one of the ingredients in the pursuit of world-class quality.

Chapter Two

Faster, Better, Cheaper: Design Engineering

Lance A. Ealey

There's a legendary comment concerning Rolls-Royce that was made during the long reign of the incredibly durable and beautiful automobile that became known as the Silver Ghost, the car that first earned Rolls-Royce its "best car in the world" sobriquet. The remark was made by a competing luxury-car builder, and it wasn't meant as a compliment. The Rolls-Royce, he said, is a triumph of craftsmanship over design.

The first Rolls-Royce was designed and built by the English engineer Sir Frederick Henry Royce. Royce's approach to car building was to "engineer" his designs to perfection, no matter what the cost. That's not to take anything away from Rolls-Royce: For the few who could afford them, Silver Ghosts were, indeed, the best cars in the world. Unfortunately, buying a Rolls-Royce just

---

Lance A. Ealey is a consultant with McKinsey and Company, Cleveland, Ohio.

wasn't an option for most duffers of that time (or even our own).

If you substitute the word *engineering* for *craftsmanship,* the disgruntled competitor's comment concerning Rolls-Royce points to one of the major differences in the way Japanese and Western automakers and, indeed, manufacturers in general, approach the task of designing and building products. Instead of *designing* products right the first time, many American industries have fallen into the time-consuming and expensive habit of *engineering* nonrobust designs up to required performance levels.

DESIGNS ON SALE

Contrary to popular belief, manufacturers today *aren't selling products: they're selling product designs.* When you get right down to it, the real trench warfare in the marketplace is being waged in design studios, not on assembly lines or in retail markets. A good design works for the company—it's easy to build, it's cost-effective, and it performs in the marketplace. A bad design, on the other hand, is your competition's best friend—stealing downstream-engineering cost and effort, ensuring a high number of product start-up problems, and generally making life both costly and miserable for everyone, from engineering to manufacturing to sales. Product design impacts every downstream operation in a company. If a company's designs aren't up to snuff—if they're not robust, if they're still green—there's little you can do to compensate for it. The company won't be able to offer the product at a competitive price and still be assured that the product will perform in the marketplace with a minimum of warranty costs or loss of goodwill. Your competitors from Japan, on the other hand, are taking great pains to optimize their designs.

WHAT TAGUCHI OFFERS

For an example of the power of parameter design, let's look at an imaginary product-design cycle for competing US and Japanese television companies. Both companies develop new TV sets. Through exhaustive system-design work, each set has been designed with many unique features, including voice-activated channel selection. Through close examination, however, an informed observer would recognize that the American design at this point clearly holds the advantage in terms of innovation.

After system design is complete, the Americans set about engineering the various circuits so that they conform to the given specifications. In order to maintain a high-quality picture, the American engineers set certain critical tolerances very tightly. To achieve these tolerances, high-quality components are specified, which significantly raises the consumer price for the TV.

The Japanese, after coming up with a system design, go through the parameter-design stage. During this stage, they optimize the performance of various aspects of the TV using the cheapest possible components. Because they've optimized the various circuits during parameter design, the need to specify very tight tolerances during the tolerance-design stage is greatly reduced. The Japanese utilize Taguchi's experimental-design techniques during both parameter and tolerance design to quickly and cost-effectively determine design and tolerance levels for the TV. They also discover that they can replace certain design elements with cheaper components without affecting picture quality or set reliability.

Both TVs are introduced to the market. The Japanese TV comes to market some three months earlier than the American TV and quickly gains a reputation as a high-quality/high-value unit. The American TV is late because various circuits had to be redesigned that were adversely affected by various ambient-temperature and humidity levels discovered during last-minute prototype testing. Cost overruns in several areas, including unanticipated production problems occurring with the voice-activated channel selector, have forced the American manufacturer to drop several of the innovative sales points from his TV—including the channel selector. The TV is quickly redesigned to feature a conventional channel tuner. The American manager is shocked to see that the Japanese TV—with voice-activated channel selector—is selling for less than the projected price of his TV. He orders his engineers to reduce the cost of the TV by respecing certain noncritical circuits using cheaper components. After six months on the market, the American manufacturer is inundated with customer complaints and warranty costs because the TV's performance and reliability, when compared with its Japanese competition, aren't even in the same ballpark. Many of these warranty costs, incidentally, can be traced back to the cheaper components that were substituted at the last minute for more costly, higher-quality parts.

Many industries, including the bulk of the television-building industry, have moved out of the United States. The reason typically given for moving offshore is that the company can't compete because of the cost of the US labor force. When the Japanese yen gained in value by more than 80 percent in the mid-1980s, everyone in the United States expected the American trade deficit with Japan to shrink considerably. In many cases, US products had become as inexpensive as Japan's, the reasoning went, so Americans should start buying US-made alternatives. Instead, the US trade deficit with Japan actually continued to grow for some time. Why? The Japanese have proven that consumers are willing to buy a quality product even if they must pay a premium price for it. Reducing labor costs is only a small part of the competitive equation. It's the easy part, too; that's why so many companies have hurried offshore with their products. But if low labor costs are the only thing an offshore venture is bringing to the party, its long-term competitive ability in a sophisticated market such as the United States will be a very "iffy" thing.

The more difficult, although ultimately more lucrative, route to becoming competitive starts with design.

USING IT BECAUSE IT WORKS

Dr Genichi Taguchi's Quality Engineering techniques put maximum analytical power in the hands of design engineers. In fact, Taguchi Methods were custom-designed for engineers by a brilliant engineer, and an important ingredient of their success is the special knowledge an engineer has of a product or process. This knowledge is what makes it possible to shortcut traditional test-every-combination experiments.

In a very real sense, your successes with Taguchi Methods will be tied directly to the savvy of your engineers.

Taguchi Methods allow an engineer to organize his thinking and to deal exclusively with facts. They provide a road map that leads an engineer to the optimum levels of performance a product or process can achieve without adding cost.

Traditionally, as noted above, finding the optimum level of performance for a product or process has been something of a crapshoot in American industry. A product would be designed to meet certain artificially set performance criteria. Specification limits for a product are usually set according to inherited knowledge of that product—in other words, if it worked on the last version, it'll work on this one, too. Your engineers would eventually get you close to your target, often by specing closer-than-necessary tolerances and higher-performance/higher-cost critical components. Many times, that higher-cost component could have been replaced by a lower-cost device that had been optimized to its fullest.

The analogy of looking for a needle in a haystack is a good one in describing traditional product- and process-design techniques. There can be literally thousands of ways that a complex product can be put together. Testing every combination of variables is impossible in a manufacturing environment because problems tend to be fairly complex and time is almost always in short supply. Some combinations will give you optimum results, but not for the minimum cost. Traditionally, unless an engineer performs a full-blown design of experiments, testing every combination of factors and interactions through an almost endless series of experiments, he has no way of knowing if the combination of design factors he chose while designing a product was the best possible combination.

Taguchi's methodology allows the design engineer to shortcut that long, drawn-out procedure, and it allows him to do it without the aid of a trained statistician.

Training

An engineer can usually grasp enough of the fundamentals of Taguchi's off-line quality-control techniques to begin performing process-related design experiments after a week's training in Taguchi's Methods. Be aware, however, that it takes years for an engineer to become an expert.

Within many Japanese companies, training in Taguchi Methods is considered a necessary and continuing part of every engineer's education. A Japanese engineer without knowledge of Taguchi Methods is said to be only half an engineer. Most of this training is done in-house and is augmented by regularly scheduled symposia during which particularly interesting and successful case studies are presented. (It's been Taguchi's philosophy that the only way to truly understand his methodology is by participating in Quality Engineering case studies.)

Within the United States, Ford is a recognized leader in promoting Taguchi Methods training—not only for its own engineers but for those of supplier companies as well. The ITT Automotive Group also has an aggressive in-house Taguchi Methods training program. AT&T Bell Laboratories is another leader in in-house training, as is Xerox Corp. Still, compared with leading Japanese companies, the Americans are just beginning to climb the steepest part of the Taguchi Methods learning curve. It has been estimated that many Japanese engineers complete as many as 600 hours of training in Taguchi Methods.

Nippondenso Co., Ltd., for example, despite having more than 30 years of experience in Taguchi Methods, retains a very aggressive Taguchi Methods training schedule. About 15 percent of Nippondenso's 6,000-person engineering staff is considered expert in the methods, and another 30 percent is very competent.

The rest of the engineering staff is taught and advised by 918 expert practitioners. By comparison, only a relative handful of the engineers in most leading American companies that have embraced Taguchi Methods have even gone through preliminary Taguchi Methods training. And the number of experts in Taguchi Methods in the entire United States can literally be counted on one's fingers.

At Nippondenso, all new engineering employees go through a four-day introductory course in experimental design. Nippondenso trains about 400 people per year in this course. Within three to seven years of employment, engineers are given an additional 12-day course in the use of various Taguchi Methods techniques, including orthogonal arrays and the QLF.

Within five to nine years of employment, engineers go through an advanced 11-day session of Taguchi Methods. Additionally, reliability engineers take a separate course in parameter design and tolerance design, and engineering assistant managers attend a seven-day session on the entire system.[1]

An important part of the training is to have all trainees produce case studies for critique. Nippondenso, however, is something of a special case; it's one of the premier users of Taguchi Methods in Japan and its training program reflects this fact.

A more typical example of experimental-design training comes from the Sharp Corp., which introduced Taguchi Methods 10 years ago. Sharp provides roughly nine days of Taguchi Methods training a year for its engineering staff. It also requires engineers to attend yearly design-of-experiments seminars, which are sponsored by the Japanese Standards Association and taught by Taguchi himself. Taguchi is often retained to teach his methodology at Japanese companies. Apart from formal training sessions, the bulk of real understanding concerning Taguchi Methods comes as engineers use the techniques in the course of their daily work, aided by more-experienced practitioners.[2]

Theory, Computer Simulation, and Taguchi Methods

The traditional approach when undertaking a Taguchi Methods design of experiments is to lay out the experiments using an orthogonal array,[3] perform each experiment, and then record and analyze the data. However, the advent of accessible computers allows an experimenter to perform certain parameter- and tolerance-design case studies on a theoretical level. In situations where scientific theory exists for the actions and reactions of factors, it's possible to "perform" orthogonal-array experiments totally within the circuits of a computer. Taguchi has long advocated the use of theory in his experimental-design philosophy, since it's often more cost-effective to perform experiments on paper or in a computer than it is to physically conduct experiments.

Engineers at General Motors Corp.'s Detroit Diesel Allison Division (which has now been sold) performed just such an experimental design in developing a theoretical diesel-engine design. While the program was given a proposed budget of $10,000, the entire system was run on computer for less than $400.[4]

The engineers wanted to develop a theoretical engine of the future that would identify which combination of engine-design features would generate optimum levels of fuel efficiency.

The engineers used a computer simulation program that acted in effect as their "engine." Six factors were to be evaluated and in each case the engine's goal was to generate 300 horsepower at an engine speed of 1,800 revolutions per minute. The engineers also suspected that there were three interactions between three factors that could have significant effects on the performance outcome of the engine.

The engineers assigned the six main factors to a 27-experiment (L_{27}) orthogonal array and ran the experiments on their computer-simulated engine. Each factor was to be tested at three different operating levels. A little math will show that to perform a full factorial experimental design, during which every possible combination of the six factors at three levels is tested, GM would have had to run 729 experiments. The engineers came to the conclusion that even using a high-speed computer, a computer simulation of a full-factorial design of experiments would take too long to accomplish.

And if GM had elected to perform 27 experiments involving structural changes to the engine, the company would have had to construct the equivalent of 27 different engine configurations. While engineers will usually use one-cylinder test engines that can be reconfigured to replicate the 27 combinations of factors, the time and money spent to do so would be many times as expensive as the computer simulation.

After running the experiments, the GM engineers analyzed the data using a commercial personal-computer software program. This program is custom-tailored for Taguchi Methods experimental-design analysis, and it does most of the tedious and exacting number crunching, thus reducing the chances of calculation error.

The optimum levels of the six factors were revealed during analysis, and a confirmation run using these factor levels proved the superior fuel efficiency of this engine. Best of all, the engineers calculated that the simulator-driven Taguchi Methods experimental design generated a 17-to-1 return on investment based on the cost of the project and the man-months needed to complete it.

Intelligent Orthogonal Arrays

Today, it's possible to perform parameter-design experiments via computer simulation, removing much cost and drudgery from the experimental-design process. But what about tomorrow?

Just as computers themselves have become more user-friendly over the years, recent development work at Bell Labs indicates that the full power of Taguchi's experimental-design system could soon become even more accessible to engineers.

Madhav Phadke and two former colleagues, Newton Lee and Rajiv Keny, have developed an expert system for experimental-design techniques that automates the selection of an orthogonal array and the design of an experiment. The Prolog-based computer-software system uses artificial intelligence to assist less-than-expert practitioners in the design of complex experimentation.[5]

The prototype system, which was developed to assist AT&T engineers in running Taguchi Methods experiments, asks the experimenter a few questions concerning the experiment to be planned and then suggests an appropriate orthogonal-array design that could be used for the experiment.

There are 17 basic orthogonal arrays that engineers generally choose from. These 17 arrays run from the L_4, which will accommodate three two-level factors arranged in four different experiments, to the L_{81}, which will accommodate 40 three-level factors arranged in 81 different experiments.

While in many basic experiments the selection of an orthogonal array can be fairly clear-cut, some applications can be quite complicated. For example, in several arrays that can accommodate only two-level factors, it's possible to combine two columns of two-level factors, thereby creating the equivalent of a four-level factors column. There are also certain arrays that are best suited for dealing with the main effects of control factors.

Bell Labs' expert system will also assign the factors to specific columns in the orthogonal array using a computerized version of Taguchi's linear graphs. A linear graph is a one-dimensional depiction of the relationships between columns in the orthogonal array. Every column in the orthogonal array is represented by either a dot or a line drawn between two dots. The columns represented by lines indicate that those columns may be used to observe the interactions between the two dots (columns) they join. While a limited number of linear graphs are usually used for each array, an experimenter facing unique requirements may wish to tap more unusual linear-graph arrangements. The number of graphs available for certain arrays can be formidable.

The expert system is able to handle complex or simple linear graphs equally well and commits no mental errors or silly mistakes. As it stands now, Bell Labs' software can solve roughly 85 percent of the real-world experimental-design problems it would face in the field.

Future goals include teaching the system tricks, such as combining noninteractive factors on a single orthogonal-array column so that a smaller array could be used, which would reduce the number (and cost) of experiments.

Another area under study involves problem solving. For example, the system could suggest to the user that by dropping noncritical factors, the experiment could run on a smaller orthogonal array—again, with cost-cutting in mind.

In addition to other refinements, the software designers are exploring the possibility of giving the system an explanation capability. This would allow the software to explain why it made certain decisions and recommendations in the layout of the experiment, thereby functioning as a teaching element. At present, AT&T has no plans to sell the system outside of the company, but such research points to the day when any moderately trained practitioner will be able to tap into expertise concerning Taguchi Methods any time a need arises.

APPLYING TAGUCHI METHODS

Indiscriminate use of any quality tool can produce less-than-optimum results. As Taguchi himself has said, if you can obtain the same results by changing one factor at a time and instead you use experimental-design techniques, you're wasting both time and money.

So how do you know when and where to use Taguchi Methods? Many cases will actually shout for their use. Flex Technologies, Inc., a small automotive supplier located in Midvale, Ohio, faced a problem involving post-extrusion shrinkage of the plastic speedometer casings it manufactured for GM. The casings had been in production for more than 15 years in total at one of GM's captive-supplier divisions or at Flex Technologies. The GM captive supplier had spent a great deal of money trying to solve the shrinkage problem by changing one process factor at a time, but with little positive result.

The excessive shrinkage of the speedometer casing could cause the assembly to be noisy under certain conditions. The engineers at Flex Technologies, headed by Jim Quinlan, who had received training in Taguchi Methods at the American Supplier Institute, decided to run a design of experiments to reduce the shrinkage. In choosing which factors they should experiment with, Quinlan asked for opinions from everyone, including customers, production technicians, quality people, and engineers. From this information, Quinlan's team developed a cause-and-effect diagram for the experiments, listing all the possible causes for the shrinkage problem.[6]

Instead of performing the 32,768 experiments that would have been necessary in a full-factorial experimental design, the Flex Technologies engineers assigned the 15 factors to a 16-experiment (L_{16}) orthogonal array and ran the experiments, after which four samples of the casing produced during each experiment were subjected to heat soaking to simulate the percentage of shrinkage that each experiment produced.

By applying Taguchi's Signal-to-Noise Ratio, the experimenters discovered that eight of the factors were having a significant effect on the shrinkage. By adjusting four of these, some of which meant changing basic design specifications, the engineers were able to reduce the amount of shrinkage dramatically, from 0.25 to 0.05 percent.

Using the QLF based on the $80 warranty cost the manufacturer would incur to replace the cable casing, Quinlan was able to document a cost per unit that dropped from $2.12 to $.013. This was a clear-cut example of a situation for Taguchi's techniques.

Not all cases are so obvious, however. It's relatively easy for a manufacturer to confuse the magnitude of a problem with the importance of solving that problem. For example, which is more important: solving a reject problem on a plastic part that produces noticeable amounts of scrap, but which actually costs the company a relatively modest amount or solving a product-design problem that would not be obvious to anyone on the shop floor, but which would cost the company a fortune through a combination of downstream rc-engineering and process redesigns? The largest cash drains are usually the most invisible when viewed on a day-to-day basis. They usually fall through the cracks in an organization—until they start registering in the profit-loss column.

Likewise, it would be unwise for a company to spend large amounts of money, time, and effort improving a product's quality characteristics if the ultimate consumer of that product couldn't tell the difference or possibly wouldn't even care. But how does a manufacturer spot which quality characteristics mean a lot to consumers and which mean relatively little?

It requires vigilant attention to the marketplace and to the needs and desires of consumers. This often requires an unusual degree of sleuthing. For example, a

Japanese automaker wanted to know just what sort of cargo Americans typically haul in the trunks of their cars. To find out, it sent engineers to Walt Disney World in Florida, where they actually watched what people put in and took out of their trunks. They noted which trunks seemed easiest to use, which seemed more difficult, and so on. Using this information and other surveys of car owners, they would make sure their car trunks included those features consumers wanted most: spaciousness, easy access, and so forth.

Many companies today are using a planning tool known as Quality Function Deployment (QFD) to focus in on the most appropriate areas in which to deploy Quality Engineering. QFD effectively translates the wants and desires of the customer into language people in the marketing, design, engineering, and manufacturing departments can understand, allowing them to work together without losing sight of their goal: delivering a product the customer truly values.

Today, the design engineer's job involves a lot more than simply sitting in front of a CAD screen designing products. The design engineer has to be in touch with the marketplace so that he or she can make advantageous use of the developing technology. The design engineer must be keenly aware of the things customers like and don't like regarding a product and why. He or she alone can seek innovations in the laboratory that could meet customer needs that even the customer isn't yet aware of. But the only way he or she can do this is through constant communication with the market.

The modern design engineer must be part product planner, part market researcher, part consumer, and part visionary.

NOTES

1. "The History of Nippondenso Activities of Application of the Taguchi Method," April 1987. Provided by Nippondenso Co. Ltd.

2. Information supplied by Sharp Corp., Osaka, Japan, April 1987.

3. An orthogonal array is a matrix of numbers arranged in rows and columns, each row representing the state of the factors and each column representing a specific factor that can be changed from experiment to experiment.

4. Jerry Roslund and Mustafa Savliwala, "A Theoretical Engine Design Using a Taguchi Factorial," presented at University of Michigan Quality Assurance Seminar, August 1986, Traverse City, Michigan.

5. Telephone interview with Madhav S. Phadke, August 1987. Also, Newton S. Lee, et al., "An Expert System for Experimental Design: Automating the Design of Orthogonal Array Experiments," *ASQC Quality Congress Transactions.* 1987.

6. Jim Quinlan and the engineering staff of Flex Technologies, Inc., "Product Improvement by Application of Taguchi Methods," 1985 Taguchi Application Award, American Supplier Institute.

Chapter Three

Taguchi's Quality Engineering Philosophy and Methodology

Shin Taguchi

Dr Genichi Taguchi has developed a collection of pragmatic methods that led to the discovery of how to reduce product and process variability. These methods represent a complete system that can be applied in research and development and product design and manufacturing. As a result, products can be rapidly developed that perform well at high quality and low cost.

Taguchi calls this system Quality Engineering, although most people outside of Japan refer to it as Taguchi Methods™. Out of respect for Taguchi, my father, I will call it Quality Engineering in this paper. This approach takes a different view of three major issues regarding product quality:

Shin Taguchi is a vice president of the American Supplier Institute, Allen Park, Michigan.

1. How to evaluate quality.
2. How to efficiently improve quality and cost.
3. How to cost-effectively monitor and maintain quality.

HOW TO EVALUATE QUALITY

A company is in business to make a profit. To maintain financial health, its products must be competitive with those of other companies. The relationship among profit, sales, and cost can be represented by the following equation: profit = sales − cost.

Profits are increased by increasing sales while decreasing costs. Quality improvement is the most effective way of achieving a simultaneous increase in sales and cost reduction. This may seem paradoxical because our first reaction is that an increase in quality automatically means an increase in cost.

Cost and quality are linked in the way in which customers evaluate products. The way in which we evaluate quality must take this relationship into account, allowing us to continuously improve product quality and reduce cost. A company improves quality in order to remain competitive. We must be able to evaluate quality in a way that reflects corporate objectives. Most Western companies use the following types of measures as indicators of product quality: number of defects, percent defective, number of failures, MTBF, scrap/rework cost, yield, process capability indexes, warranty and service information, competitive analysis, and so on.

These indicators are artifacts that detect poor quality after the fact. They are cost-related and tell us that action is needed; therefore, they are important for management to monitor. From a technical standpoint, they are poor indexes to use for assessment of activities to develop or improve quality. In reality, these are only symptoms of poor product or manufacturing process functions or of the intended function's variability. For example, it is not the function of an automobile brake to squeal. It is not efficient or effective for engineers to measure squeal in order to reduce it.

A useful evaluator of quality for quality improvement activities by engineers must satisfy the following criteria:

1. It can be used in upstream quality improvement. Many of the usual measures of quality rely on post-design or post-production figures, available only after the product has been designed and tested or manufactured. By that time, it is too late to improve quality. We must be able to evaluate quality in upstream stages—in research and development and product and process design.

2. It must appropriately express customer and engineering objectives. Our measures of quality must truly express the quality of the product/process in terms of engineering intent and customer response to product function. It must be able to evaluate quality in terms of ideal product function.

3. It must express the relationship between cost and quality. It must include the effect of costs that arise because products fail to function as expected in the hands of the customer. These costs affect market share and long-term profit.

Taguchi has proposed two evaluators of quality that effectively meet these criteria: the Quality Loss Function (QLF) and the Signal-to-Noise (S/N) Ratio.

Every product, process, or service has a specific function and an ideal performance target. This represents the engineering intent of the product to which customers will respond. Quality Engineering focuses on reducing variability around the ideal function; the smaller the variability, the higher the quality. Both the QLF and the S/N Ratio are measures of this variability.

The QLF converts this variability into monetary terms that represent an estimate of the costs a customer incurs when a product fails to function as intended. These costs can include scrap/rework costs, replacement costs for failed products, time and money spent for repair, and so on. Taguchi calls these costs loss (i.e., loss to society). The QLF provides a quantitative estimate of this loss.

Thus, the QLF evaluates quality in terms of ideal product function and customer response. It can be used in the earliest stages of product planning and development, as well as in production. Because it evaluates quality in monetary terms, it is used to effectively solve cost–quality trade-off problems. This is done by finding the balance between cost and quality that can increase profit. Cost-effectively improving quality reduces loss. In management terms, the objective of quality improvement is to reduce loss.

The S/N Ratio also measures variability around target performance. It is a measure of the stability and reliability of performance in the face of those uncontrollable factors (i.e., causes of variation) that a product will encounter on the factory floor and when used by the customer. The S/N Ratio gives the design engineer a reliable upstream estimate of how a product will function at downstream stages. It evaluates quality in terms of decibels (dB) and serves as an ideal characteristic for measuring quality improvement in engineering terms. In engineering terms, the objective of quality improvement is to increase the S/N Ratio.

Whatever the system (material, product, manufacturing process, etc.), an "intent" can be identified. The intent is what the system is designed to do. The intent of an automotive brake, for example, is to smoothly slow down the vehicle. The intent of an injection molding machine is to create the intended shape (dimensions) with a certain property. In order to achieve the intent, any engineering system uses some form of energy transformation. Assuming the intent is what the customer wants, any quality problem is a symptom of energy-transformation variability.

If we are interested in developing a high-quality system, it is far more efficient and effective to reduce energy transformation variability by maximizing S/N rather than reducing problems by measuring symptoms.

A typical design process is to repeat design-test-build activities. First, the product/process concept is developed and a prototype is built. Next, the prototype undergoes testing to find problems by observing the symptoms (i.e., problems). Then, it is redesigned in an attempt to reduce problems. And then it is tested again.

Redesign-build-test-redesign-build-test—this process will be repeated until the deadline occurs or the design is acceptable. With this approach, we usually are not evaluating the intended function's variability. Rather, we are measuring symptoms. Most typically, solving one problem will introduce another.

Conceptually, the S/N Ratio is represented by:

S   Energy transformed to perform the intended function.

N   Energy transformed to other than the intended function.

FIGURE 3–1
The Concept of Signal-to-Noise Ratio.

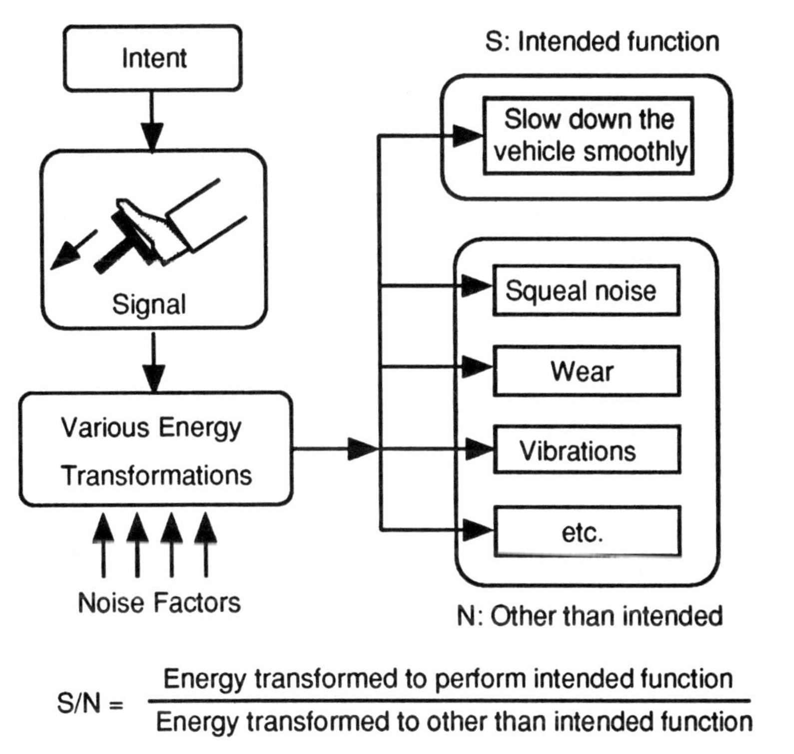

$$S/N = \frac{\text{Energy transformed to perform intended function}}{\text{Energy transformed to other than intended function}}$$

Figure 3–1 illustrates this, using a braking system as an example. Squeal noise is a symptom of a poor brake. It is not the intended function of a brake to have squeal noise. We may be able to reduce squeal noise by measuring it (not easy because of strong interactions among control factors), but it may not necessarily improve the S/N Ratio. It may worsen other symptoms (or failure modes) or even introduce a new problem. In order to reduce squeal noise, what should be measured?

The theme of the 1989 Taguchi Methods Symposium was "If you want quality, don't measure quality." This thinking is very important for the productivity of engineering activities. The function of the braking system is to create torque to generate deceleration. We should measure the relationship between the foot force (the signal) and the generated torque (the output response) and then evaluate the relationship's variability using S/N. This relationship and the concept behind the calculation of S/N are shown in Figure 3–2.

Variability causes various symptoms, such as squeal and vibration. Increasing S/N by 6 dB is equivalent to reducing variability by half. Moreover, the QLF and S/N Ratio are related measures. They are both measures of variability around the intended function. We can easily convert the S/N Ratio into the QLF and thus evaluate quality improvement in monetary terms. An increase of S/N by 3 dB is equivalent to reducing loss by half. Engineering objectives can be expressed in terms relevant to management, and this facilitates communication and the achievement of company objectives.

HOW TO COST-EFFECTIVELY IMPROVE QUALITY (OFF-LINE QUALITY CONTROL)

Three Stages in Product/Process Development

Quality Engineering is a system for achieving corporate engineering objectives (i.e., reducing loss and improving profit). Off-line (design stage) quality improvement includes the following three stages:

1. System design—conceptual design.

FIGURE 3–2
Calculation of Signal-to-Noise Ratio.

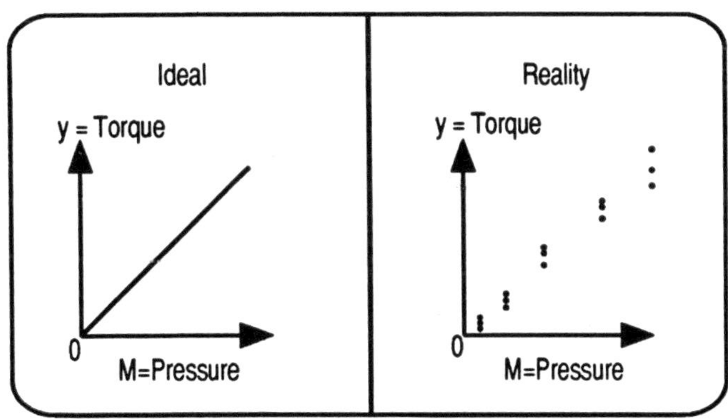

Automotive Brake

M = Signal = Pressure

y = Output Response = Torque

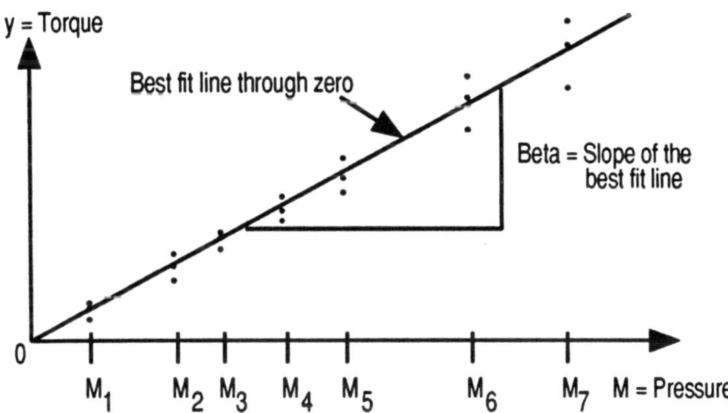

$$\text{S/N} = \frac{\text{Power of proportionality between M and y}}{\text{Variability around the proportionality}}$$

$$= 10\log\frac{\text{Beta}^2}{\text{Sigma}^2}\ (\text{dB})$$

Note: 1. Sigma^2 is the average of square of distances from individual points to the best fit line.

      2. Notice that the higher the S/N, the better the ability to stop the car smoothly.

2. Parameter design—robust design without a cost increase.
3. Tolerance design—cost-quality trade-off.

System design is the conceptual design stage, in which scientific and engineering expertise is applied to develop new and original technologies. This involves creativity. For example, the original conceptual design of the video cassette recorder (VCR) was developed in the United States.

Quality Engineering techniques do not focus on this stage. Since it is not possible to study all potential systems (unless computer simulations are performed), Taguchi suggests that engineers select one, or a few, concepts for development. A generic Quality Function Deployment (QFD) study or Pugh Concept Selection can be used to make the selection.

Parameter design is the stage at which a selected concept is optimized (quality improved or S/N maximized without a cost increase). Many variables can affect system function. These need to be categorized from an engineering point of view. Some will be controllable (control factors). The engineer is free to set the nominal values of these variables. Others will be uncontrollable or too expensive to control (noise factors). These are the downstream variables that cause functional variation (see Figure 3–3).

The goal of parameter design is not to find and remove these causes of variation or to upgrade material and components to improve quality, but to find that combination of control factor settings that allows the system to achieve its ideal function and remain insensitive to noise factors.

This allows us to develop designs with high performance, stability, and reliability. This stability of perfor-

FIGURE 3–3
The Engineered System and Parameter Design.

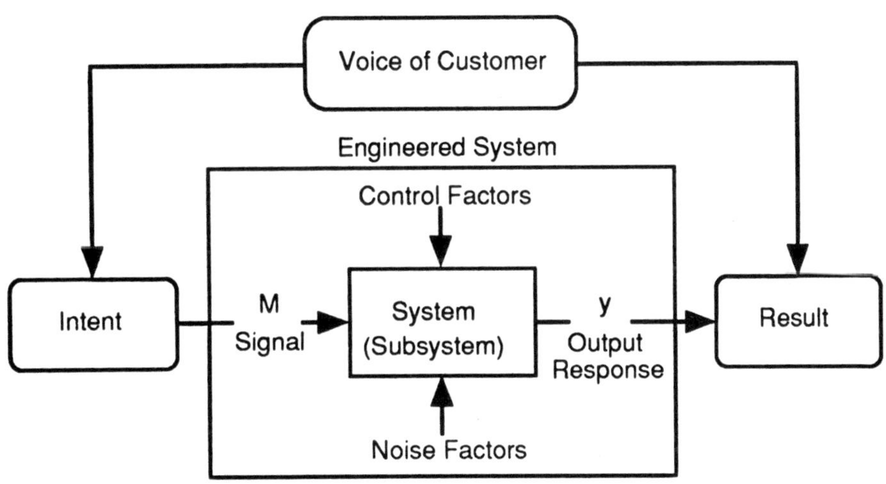

Parameter Design (Robust Design)
1. Design the experiment with Control Factors
2. Measure Output Response (y) as Signal (M) and selected noise factors are varied
3. Calculate S/N and find the combination of Control Factors that maximize the S/N

mance in the face of noise factors is called robustness. Improving robustness through parameter design allows us to improve quality without increasing cost by designing products that are highly reliable under a wide range of conditions, using inexpensive materials and parts.

The S/N Ratio evaluates the robustness of the function at the selected control factor levels. The objective of parameter design is to reduce loss by increasing robustness.

There is another benefit of parameter design in addition to quality improvement, however. Parameter design provides opportunities to reduce the product and manufacturing cost. Once we can improve quality without increasing cost, we can afford to apply various cost-reduction measures.

Suppose S/N was increased by 20 dB without a cost increase. We can then apply some cost-reduction measures, which may result in reducing S/N by 10 dB. However, we can afford to do so because the quality is still 10dB better than the original condition, and the cost is simultaneously less. Therefore, we can increase sales while reducing cost. This increases a company's ability to apply technology to compete economically (corporate technological capability).

Tolerance design is the stage at which variability can be further reduced by tightening tolerances. This requires the upgrading of a system's materials, parts, and subcomponents and generally increases cost. Tolerance design attempts to further reduce variability with a minimum cost increase. It uses improvement measures for controlling some noise factors and/or compensation for the effects of noise factors, but it does this rationally.

Rather than a costly upgrade of everything, tolerance design uses the QLF to evaluate the impact of improvement measures with respect to cost, and a trade-off between cost and quality is found, balancing the two.

Even when parameter design results in sufficient variability reduction, it is important to conduct tolerance design to determine tolerances and grades. When more robustness is achieved via parameter design, less upgrading is needed in tolerance design or perhaps none at all.

Selection of Characteristics (Determination of S/N Ratio), the Product/Process Engineer's Most Important Task

As previously mentioned, the selection of quality characteristics that properly reflect a product or process's engineering function is the most important, and perhaps the most difficult task of the quality engineer. To evaluate robustness or improve existing product/process performance, what should we measure in order to accurately express the data and efficiently reduce variability (i.e., maximize S/N)?

Characteristics such as yield, number of defects, percent defective, probability of failure, product life, MTBF, categorical sensory evaluation, and so on, are often used. Such characteristics do relate to economics and are needed to monitor and confirm product and process quality. However, such characteristics do not directly express engineering function. They are often quite removed from the fundamental physics of product function and involve numerous extraneous interactions that can lead to problems downstream. Using these characteristics can compromise the efficiency and reliability of the experimentation.

For the upstream evaluation and improvement of products and processes, we need characteristics that accurately reflect engineering intent—those that are close to the basic energy transfer of the system and can be used to express performance, reliability, and durability.

For example, a painting process may produce various types of defects, such as voids, sags, and so on. Rather than counting the number of such defects, or measuring yield as an output characteristic, a better strategy would be to recognize the painting process's intended function. Then we may decide to measure the relationship between the paint material flow rate (the signal) and the paint thickness (the output response). Our objective would be to reduce the variability of the relationship (i.e., achieve a robust relationship)—and create a process that can produce any thickness with the smallest variability.

This allows us to better study the variability of energy transfer without as many intervening variables. The use of such data as occurrence of defects, or yield, for reducing variability in the system is an inefficient strategy. Yield is an ineffective measurement because the sources of variability are compounded and due to interactions of the symptoms of poor robustness.

Occurrence of defects is an inefficient measurement because it measures the symptoms of poor robustness and only leads to fire fighting or problem solving, without improving the system's robustness. The selection of the best quality characteristic to express product/process function is case-dependent and depends upon the engineer's knowledge of the particular field of technology.

Design of Experiments

Quality Engineering uses designed experiments to improve S/N of products and processes. The objectives of experimentation in Quality Engineering are significantly different from those of traditional Design of Experiments.

Traditional experimentation is geared more toward meeting the objectives of the research scientist, who seeks to discover and describe natural law or phenomena. The objective of the corporate engineer is to achieve ideal product function to reduce variability at the lowest cost and in the shortest development time. In learning to do this, the engineer accumulates technical knowledge for his or her company. He or she needs methods to efficiently acquire technical information for improving quality while reducing cost. The engineer's activity is constrained by cost, time, and resources.

The engineer's job is to specify product/process controllable parameters' nominal values, their tolerances, grades of materials, components, and so on, in such a way that cost and quality can be optimized. In addition, he or she must do this as quickly and efficiently as possible. The engineer must design products that reduce loss.

In parameter design, an orthogonal array (experimental design matrix) should be used for controllable parameters (control factors), as shown in Figure 3–3. The experiment is design in terms of control factors. For each combination of control factors, the output response is measured as signal and selected noises are varied so an S/N Ratio can be calculated. Then, each control factor level is evaluated in terms of S/N. The most robust design is where S/N is maximized. Next, those control factor levels that maximize S/N are selected for use.

The objective is to achieve a robust function of the system, not necessarily to understand the mechanics of a phenomenon. While Quality Engineering uses a balanced (orthogonal) experiment, as in traditional Design of Experiments, the application of these techniques and the philosophy that underlies them were developed for use by engineers to fit the realities of industrial design and production.

HOW TO COST-EFFECTIVELY MONITOR AND MAINTAIN QUALITY (ON-LINE QUALITY CONTROL)

Even after quality has been improved and costs reduced using off-line methods, sources of variability still remain at the production level. These include variability in materials and purchases, components, tool wear, machine fatigue, drifting of process parameters, measurements of error, and so on. To maintain the benefits of off-line improvements, these sources of process variability must be controlled or compensated.

Taguchi has developed a system of on-line quality control methods that utilizes the QLF. It allows manufacturing engineers to design process control systems that minimize loss by monitoring the balance between cost and quality.

An on-line reduction in variability comes from the measurement and adjustment of equipment and process parameter levels. These procedures have associated costs, but Taguchi's on-line system takes them into account. It determines how often a product characteristic or a process parameter should be measured and adjusted and what the optimum adjustment limits should be. This allows for economically informed decisions regarding production control measures, in order to optimize process control systems.

This is quite different from the widespread strategy of control charting. Feedback and feedforward control are very important in flexible manufacturing and the design of automated processes. Automated control systems can produce expensive losses if they are not optimized.

Feedback control measures a product or process parameter every n units and adjusts the process closer to the target using an adjustment factor. It thus compensates for the effects of variation rather than trying to remove the variation's causes. The frequency of measurement and the control limits are optimized as a function of the costs of measurement, adjustment and scrap/rework, the process's capability, and manufacturing tolerances.

Taguchi's on-line methods can determine how often to measure and what the control limits should be. His on-line system includes methods for product control, prediction and correction, diagnosis and adjustment, and preventive maintenance. What to measure, what to adjust, how much to adjust, how to improve adjustability, and how to improve measurement are dealt with through off-line methods.

By the application of off-line and on-line methods, control can be improved, monitored, and maintained on the basis of an optimum balance between cost and quality.

Chapter Four

Educating Today's Engineers

Robert H. Schaefer

Moving from a design-development process that's deterministic in nature to one that's statistically based is a journey our engineers must make. Approximately two years ago, the Chevrolet-Pontiac-Canada (C-P-C) Group of General Motors Corp. decided to embrace the concepts of Dr Genichi Taguchi and started formalizing plans for the implementation of his approach. Rather than dream up a strategy, we asked Dr Taguchi for his advice on implementing these methods at C-P-C.

Dr Taguchi said that we must do two things: use the case-study process and develop resident consultants. We did what he said and it appears to be working very well. I'd like to share with you the details of our strategy for implementing the Taguchi approach and the roadblocks and cultural issues we encountered and how we handled them.

Our implementation strategy isn't brilliant; as a matter of fact, it's fairly straightforward. It's basically a three-

Robert H. Schaefer is director of systems engineering methods, General Motors Corp., Warren, Michigan.

pronged approach: the team case-study process, the resident Taguchi consultants, and the Taguchi Council.

Let's begin with the team case-study process. Taguchi Methods are taught through team training. Our teams are cross-functional and generally consist of five or more people. Most often, the teams are made up of engineers, manufacturing engineers, and engineers from our suppliers. But the teams aren't limited to engineers. We also include technicians and anyone else who can contribute to the project. Each team is required to bring a qualified real-life project to class. Team members receive 40 hours of training over several weeks, which gives them time between classes to gather data and work on the project. A resident Taguchi consultant is assigned to each team to help it through the rough spots. We try to make each case study a success.

Dr Taguchi also told us that management must facilitate the case-study process. What that means is that management must participate in discussions with the teams and recognize them for their efforts. Bob Schultz, C-P-C Group vice president, and his staff review one case study each month. This provides several benefits. It's an excellent way for top management to learn the benefits of the Taguchi approach, and staff members learn the language and learn to ask the right questions. It's also an excellent way to recognize the team, and it lets the work force know that top management is interested.

In August of 1986, C-P-C held its first symposium. Both C-P-C and corporate leaders participated. That symposium, like ASI's annual Taguchi symposium, was very motivational. It gave our case-study process a real shot in the arm. People tell us that we're lucky to have top-management support. Well, it wasn't that way in the beginning. We had to *earn* their support. How did we do it? With case studies that solved significant problems and yielded substantial savings. Several success stories were all it took.

The second element of the three-pronged approach is the resident Taguchi consultants. These consultants are the stars of earlier case studies. In addition to their regular jobs, these people work with the newly formed teams.

All of us at one time or another have gone to a class on a technical subject. After the class was over, we went back to our jobs and attempted to put into practice what we learned. What seemed so simple in class was no longer simple: It was complex and we needed help. This is precisely the time when an expert is required. Our resident Taguchi consultants help the team by questioning its approach, making suggestions, and even assisting with data analysis. We've found that the consultants' contributions are invaluable. In addition to working with the teams, the consultants perform on-the-job training with their peers. This allows us to go beyond the case-study process itself in spreading the use of the Taguchi approach.

Dr Taguchi suggested that an optimum ratio of consultants to peers is 1:20. By the end of 1987, C-P-C had 1:100, and by the end of 1989, we hope to achieve the 1:20 ratio.

Figure 4–1 shows how we track the expertise levels of our engineers. In this example, test engineer Bill Biondo is the resident Taguchi consultant for his chassis test group. He spends approximately 50 percent of his time consulting with his peers in the test community. He also helps with the case-study teams. Bill wasn't singled out or chosen for this task. He, like all the other consultants, stepped forward because of interest and requested training. That's how it starts.

FIGURE 4–1
C-P-C Uses an Expertise Summary to Track the Training Level of Its Engineers in Various Quality Control Methods. The Third Engineer on the List, Bill Biondo, Is a Resident Taguchi Methods Expert.

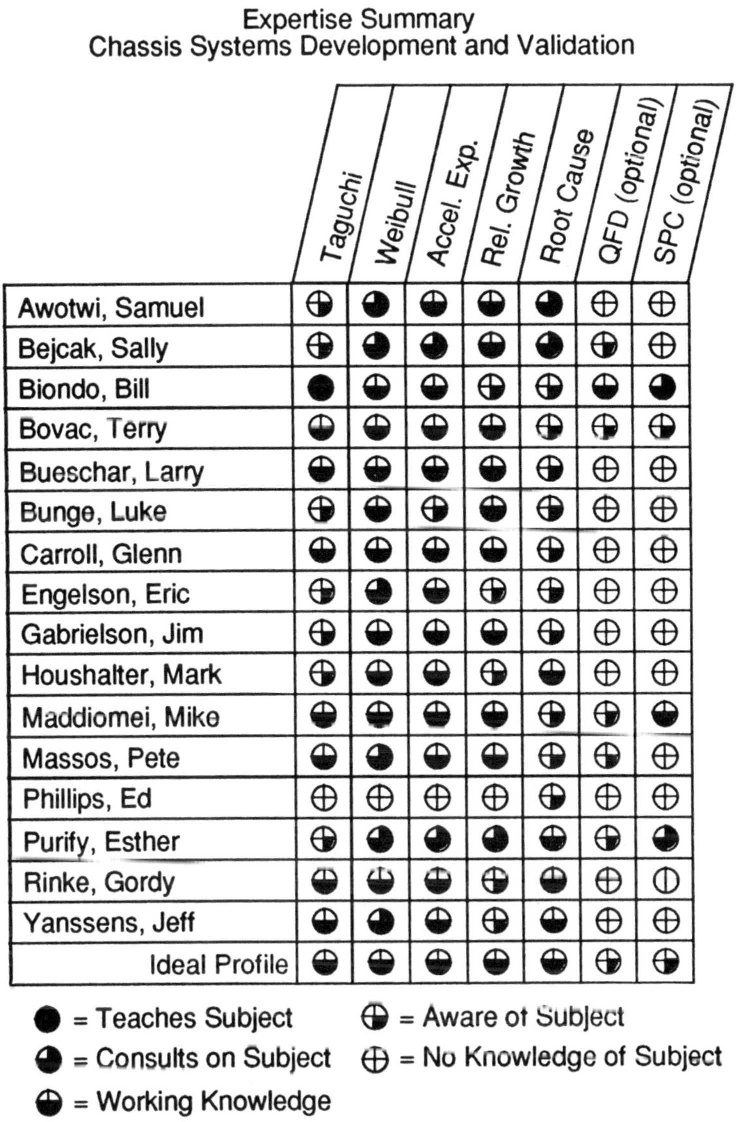

Expertise Summary
Chassis Systems Development and Validation

	Taguchi	Weibull	Accel. Exp.	Rel. Growth	Root Cause	QFD (optional)	SPC (optional)
Awotwi, Samuel	◔	◕	◒	◒	◕	⊕	⊕
Bejcak, Sally	◔	◕	◕	◒	◕	⊕	⊕
Biondo, Bill	●	◒	◒	◔	◔	◒	◕
Bovac, Terry	◒	◒	◒	◒	◔	◔	⊕
Bueschar, Larry	◒	◒	◒	◒	◔	⊕	⊕
Bunge, Luke	◔	◒	◔	◒	◔	⊕	⊕
Carroll, Glenn	◒	◒	◒	◒	◔	⊕	⊕
Engelson, Eric	◔	◕	◔	◔	◔	⊕	⊕
Gabrielson, Jim	◔	◒	◒	◒	◔	⊕	⊕
Houshalter, Mark	◔	◒	◒	◔	◒	⊕	⊕
Maddiomei, Mike	◒	◒	◒	◒	◔	◒	◕
Massos, Pete	◒	◕	◒	◒	◔	◒	⊕
Phillips, Ed	⊕	⊕	⊕	⊕	⊕	⊕	⊕
Purify, Esther	◔	◕	◕	◕	◕	◔	◕
Rinke, Gordy	◒	◒	◒	◔	◒	⊕	◑
Yanssens, Jeff	◒	◕	◒	◔	◒	⊕	⊕
Ideal Profile	◒	◒	◒	◒	◒	◔	◔

● = Teaches Subject   ◔ = Aware of Subject

◕ = Consults on Subject   ⊕ = No Knowledge of Subject

◒ = Working Knowledge

The third part of our implementation strategy is the Taguchi Council. The council is made up of people who understand and have used Taguchi Methods. These people are the true believers, the fire-in-the-belly people. They're not appointees. They come from many areas within C-P-C—from engineering to the factory floor. The council orchestrates the implementation process. It selects the teams to be trained and establishes the training criteria, which is tough. For every 10 teams that apply for training, on the average only one is chosen.

This is a hard-nosed approach. But we're interested in quality, not quantity; successes, not failures. The council also assigns the resident Taguchi consultants to the teams. Many feel that it's the dedication of this core group of people that has provided the real stimulus to our process.

And there you have it, a simple yet effective process. But what about the problems? We do indeed have our roadblocks—we're no different from anyone else. Figure 4–2, a flower picture borrowed from a bulletin board in a Japanese factory, illustrates these roadblocks. Workers at Diesel Kiki explained that this flower will grow very well on its own. All it takes is a little sunshine. Figure 4–3, the flower picture redrawn, illustrates an important point. In this figure, the sun is management and the flowers are the successful use of new methods by the work force. The sunshine the flowers require is for management to show support, ask the right questions, and provide recognition. With that, you'd think that the flowers would grow well. Not necessarily so.

Somewhere in every organization is a manager who insists on stomping on the flowers, as Figure 4–4 shows. There are many reasons for this. When things are hot,

FIGURE 4–2
This Flower Picture Borrowed from a Japanese Bulletin Board Illustrates that Flowers (Successful Use of New Methods by the Work Force) Grow Well with a Little Sunshine (Management Support).

it's difficult to free engineers for education and training—there's just too much to do. Or it's okay for my engineers to go to class, but don't ask me to stop a test

FIGURE 4–3
*If Management Shows Visible Support, Asks the
Right Questions, and Provides Recognition, the Flow-
ers (Successful Use of New Methods by the Work
Force) Should Continue to Grow.*

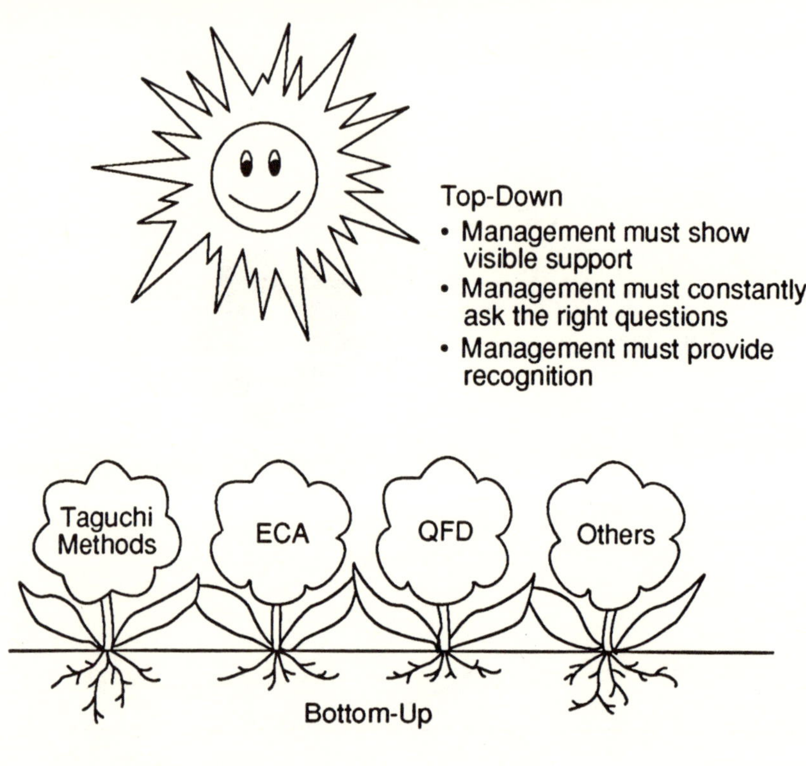

or shut down the line to get data for that case study. Or
more basic yet, "I don't want my people using that stuff
in *my* area." In addition to these three reasons, there
are at least 100 other ways to crush the flowers.

Why do managers do this? Thomas Jefferson once
asked, "Why is it that men of goodwill insist on destroy-

FIGURE 4–4
Unfortunately, Every Manager Has at Least one Manager Who Insists on Stomping on the Flowers.

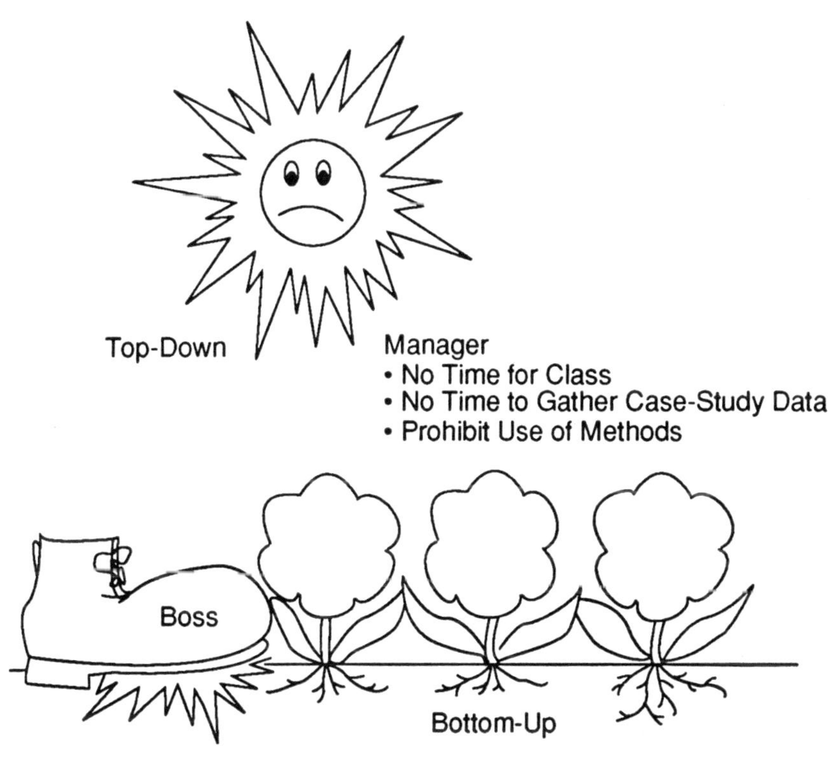

ing that which they don't understand?'' And there's our clue: If a manager doesn't understand a subject, chances are that he'll kill it. That's human nature, a fact of life. But could the reverse also be true? Our experience is that it is. If a manager understands a subject, chances are that he'll support it.

So how do we get managers to understand? Here are three suggestions:

1. Require management overviews. This is one of the few times that mandatory training is effective and necessary.

2. Have management teach. The cascade effect is a lot easier said than done. If done effectively, however, the results are extremely good.

3. Use the appraisal process to evaluate the manager on the genuine effort made to get himself and his people trained. The appraisal process is controversial, but if you have one use it for this purpose.

There are many other ideas on how to gain management support, but this should get the point across.

Of equal concern is the debating that occurs within an organization over the validity of one technical approach versus another. Debating casts a shadow and inhibits growth. Workers and management don't need this internal confusion. My advice is to take whatever steps are necessary to eliminate or minimize this problem. If left unchecked, it will undermine both efforts.

The following two points also apply to our implementation strategy. First, we believe in the "quality lever" (see Figure 4–5). Having the engineers apply the Taguchi approach during the design/development phases yields substantial return-on-investment rewards. Applying these methods during the production phase has its benefits. There's no doubt in our minds that the farther we push these methods upstream the greater the payoff. Desensitizing the design and finding potential problems during design and development are far less expensive than finding and fixing problems after tooling has been committed. We therefore emphasize the use

FIGURE 4–5
Applying Quality Technology as Far Upstream as Possible Will Provide the Greatest Payoff. Thus, C-P-C Emphasizes the Use of Taguchi Methods during Design and Development.

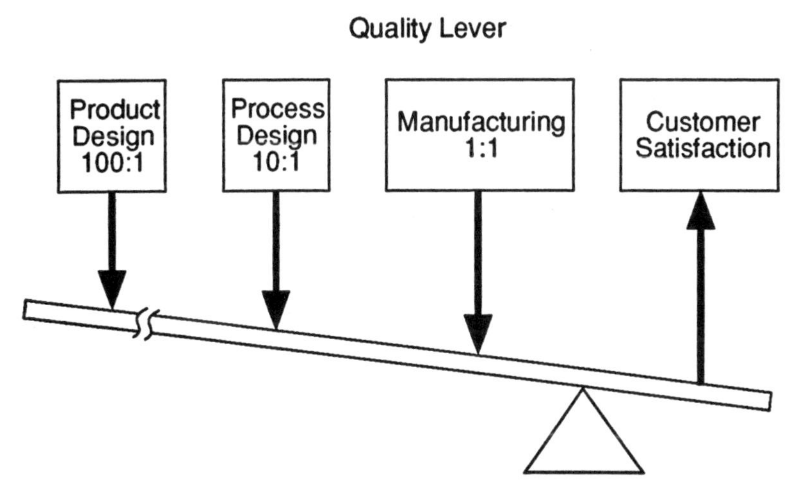

of the Taguchi approach during the design and development phases.

Second, the day-to-day jobs of our engineers are extremely demanding. The time it takes to educate and train these engineers is time taken away from the demands of their jobs, as noted earlier. C-P-C has dealt with this problem in several ways. There are those managers who understand and are 100 percent supportive. Most of these managers have had the mandatory overviews. They find the time for their people to be trained. The Taguchi Council looks to these groups for case studies. We also find that not all engineers are under the gun. For instance, the analytical and test engineers are generally more available for the application of these methods than the release engineers. By targeting our areas, we're more assured of achieving successful results. And as our successes mount, resistance crumbles.

The process described here is nearly two years old and gaining momentum. Although we're far from where we'd like to be, the process is working very well. All too often we see processes come and go. It's no surprise when the troops say, "Here's another one of those programs." But we don't hear any of that about the case-study process.

Our Taguchi Council is presently tracking 80 to 90 case studies. Recently, case studies started appearing from areas that we hadn't any idea were using the methods. If you get more than you ask for, you can feel pretty good about the process that's been set in place. Is the process self-sustaining? Not yet—we need the 1:20 ratio of resident Taguchi consultants. Then it *will* be.

Chapter Five

Specialists and Taguchi Methods

Dr Don Clausing

Editor's Note: The following essay is based on extemporaneous comments made at the Fifth Annual Taguchi Symposium, held October 8–9, 1987, in Detroit, Michigan. These comments followed the presentation made by Robert J. Schaefer of General Motors Corp., "Educating Today's Engineers."

Larry Sullivan:

How do we deal with statistical specialists and the ongoing debate that tends to paralyze management into doing nothing? This question comes up time and time again. Dr Don Clausing of the Massachusetts Institute of Technology has experienced this problem personally; he has done a lot of consulting in the industry on Taguchi Methods. He has a very good perspective on the problem, the dangers of the problem, and how to solve the problem.

Dr Don Clausing is Bernard M. Gordon Adjunct Professor of Engineering Innovation and Practice at Massachusetts Institute of Technology, Cambridge, Massachusetts. He was previously manager and then principal engineer of the advanced development activities at Xerox Corp.

Dr. Clausing:

I'd like to start out by congratulating Bob Schaefer not only for his presentation, but for the work that's gone into making the presentation possible, including all the management arrangements he described. I had the pleasure of visiting with Bob about a year and a half ago when C-P-C was still fairly new to these activities, and it's really gratifying to see the progress that's been made. In respect to the question of specialists, I think Bob already has the final answer: get rid of them. It sounds a bit draconian, but I think it's probably the only effective thing to do, particularly if people want to be argumentative, disputatious, disruptive, and not helpful. At some point you have to decide that it just isn't going to work and find some other place for them.

One of the most important of Dr Taguchi's methods is parameter design, for both product design and production processes. When I was working at Xerox Corp. in the late 1970s, we had developed our own form of parameter design. We didn't *call* it parameter design, but that's what it was. It was a method of stress testing and operating windows. In 1980, we first encountered Taguchi Methods™ through the "Blue Book" (*Introduction to Off-Line Quality Control*) by Dr Taguchi and Professor Yuin Wu. I read with great interest the first three or four chapters on the Quality Loss Function, and I said, "Boy, this is really powerful stuff; this guy Taguchi is apparently onto something here." And then I got to the Wheatstone Bridge (a type of electrical circuit) case-study example of parameter design and I couldn't understand it. There were some problems and misprints in the equations in the book, which didn't help much. So I put the book aside and said, "It looks powerful, but I can't understand it—someday I'll have to try and figure it out."

In 1982, I had the pleasure and opportunity of meeting and talking to Dr Taguchi, and I started to understand that what he called parameter design was what we were already doing. He had it much more highly developed, however, and it was much more efficient than what we were doing. But our form of parameter design had proven highly effective and made dramatic improvements over the old haphazard, one-factor-at-a-time experimentation. Once I understood that Dr Taguchi's method was better, I thought it was a really good thing; I began to implement it, and for the past five years have been helping people do the same. In a recent interview in a Japanese journal, Dr Taguchi talks about the application of Taguchi Methods and the origin of the name, for which I seem to have gotten the blame; in my more formal moments, however, I call it "Quality Engineering at Low Cost."

When we set out to implement Taguchi Methods, we found some problems—and there are a lot of problems. Bob (Schaefer) has done an excellent job of addressing them. I'll just add a few comments. One of the problems is that people, specifically engineers and statisticians, tend to react in a very specialized way; let's focus on engineers. Engineers will probably react to any new method, but in regard to Taguchi Methods, they tend to react as if they were a theorem in geometry. You remember proving a theorem in geometry: If you could show any counterexample, it proved that the whole theorem was no good and you would just throw the theorem out. That's all right for a theorem in geometry, but a lot of engineers want to use that same approach with new methods. All methods are going to have some problem in application: They're going to look like they don't work on some particular case study somewhere, sometime. If we take the Euclidean geometric-theorem-proving approach, we can just say it didn't work, so let's throw it out. If it doesn't work 100 percent every time, it can't be very good; let's not do it. To some extent, that's just a defense mechanism for the desire not to change in the first place. To the extent that it has some intellectual basis, it's an attempt to apply this Euclidean-theorem approach to methodology.

There's been some really brilliant research done on this general subject—how do we implement improvements and how do we think about them?—by Joel Moses of MIT and Rosabeth Kanter of the Harvard Business School. Moses points out that in the United States, and in the Western world in general, we tend to want methods that are general, methods that we can be confident will work every time. The problem with such methods is that they usually do so because they're not being used to do very much, and they aren't really very powerful. On the other hand, methods that are really powerful and accomplish a great deal aren't totally general. The beauty and power of them are that they work nearly all the time in situations that are really important. So a method is good if it will produce outstanding results in 95 percent of the cases. The fact that it may not produce outstanding results in the other 5 percent is okay. We have to get away from this specialized-knowledge, geometry-theorem approach.

Approached in this way, Taguchi Methods aren't going to work every time, and we shouldn't argue about the fact that there are occasions where they don't work so well. Even in those cases, we probably just need to apply them better.

Beyond that, there's the question of the role of statisticians and engineers in all of this. I'll try not to make this my usual diatribe against statisticians. Statisticians are fine people. There are, however, some problems in the way that engineers and statisticians interact. Being an engineer, I tend to attribute these problems in my more informal moments to some problem with statisticians. But it takes two in this kind of situation to have a problem. In some ways, from a historical perspective, it seems like we're repeating history.

In fact, at the turn of the century there was the new operational mathematics of Oliver Heaviside in Great Britain. Heaviside was an electrical engineer. He started doing these operational transforms and the mathematicians were all bent out of shape because the transforms were all wrong, he hadn't proven a lot of things, and so forth. But the transforms were very effective in electrical engineering. And so this battle went on for some time. Heaviside just kind of ignored it and went on successfully doing his thing, and the mathematicians went on being unhappy. Finally, some of them stopped arguing about it and started trying to prove that he was wrong, and, when they did, they found that he was right. (Heaviside later said that even Cambridge mathematicians should be given justice.)

So we have to keep it all in perspective. Even if we can't prove something 100 percent mathematically, it still may be very good; at the same time, we don't want to get too upset with people who are saying that it's no good because they haven't gotten around to proving that it's all right.

With respect to the role of statisticians and engineers, let me try to give a balanced view at this point. Another thing that comes out of the research by Moses and Kanter is that in the United States we tend to be too segmented. For example, we have engineers in one segment and statisticians in another segment, and these two segments don't talk much to each other, certainly not effectively. And it's not just engineers and statisticians: It's design engineers and production people, product planners and design engineers. We're segmented in many ways. The segmentation of statisticians and engineers, however, has a great bearing on the topic of developing better methods for optimizing products and production processes. Now as I said, at Xerox we had developed a form of parameter design. It didn't have any statistics in it, and for that reason it wasn't very efficient, but it was dramatically successful simply because the objective of parameter design is itself so powerful. We engineers had developed a powerful methodology, but it still wasn't as effective as it should have been by any means—although it was a lot more effective than the traditional poke-and-hope kind of approach.

By 1979, I had concluded that we needed to bring some statistics into the design to make it more efficient, and we started trying to do that. So we had the right objective— parameter design; i.e., reducing variance to make products work well under a wide range of circumstances. But we didn't really have the detailed statistical techniques we needed to make it more effective. The statisticians, on the other hand—and not being a statistician I can only characterize what they did with some concern—had developed a lot of techniques, but with respect to what's important in the economics and engineering of products and production processes, they really hadn't come across the right objectives. The objective of parameter design wasn't being worked on by the statistical technicians at all. I just want to leave it at that—just with the idea that the engineers and the statisticians were overly segmented.

If I as an engineer had been paying more attention to what the statisticians were doing, or if the statisticians had been paying more attention to what engineers such as myself were doing to develop parameter design—if we'd not been so segmented and had gotten together to work effectively—we'd have probably ended up where Dr. Taguchi has been for at least 10 years now. Which is to say that Dr. Taguchi didn't get boxed into these separate segmental traps; instead he did broad thinking across engineering and statistical lines and integrated these lines together to get the most effective techniques with the right objectives. We remained excessively segmented, and I'll take my part of the blame as an engineer for not looking at statistics sufficiently. And I would suggest to the statisticians that they take their part of the blame for not looking at engineering and economic objectives sufficiently and that we stop having arguments that, in the long run, aren't at all effective.

In reality, if people are going to keep on with such arguments, the only solution that effective managers can implement is the one that Bob Schaefer suggested. So I'd suggest that the real problem isn't that as an engineer I shouldn't blame the statisticians, or that the statisticians shouldn't blame the engineers; rather, it is probably fundamental to our overly-segmented culture. One group in this segment, another group in that segment; we didn't get together and do the kind of broad, integrated thinking that was needed to achieve the right techniques and the right objectives. Dr Taguchi *did,* and that's why most of us are trying to use his methods in the most effective way possible.

Chapter Six

QFD in the Service Environment

Kurt R. Hofmeister

Editor's Note: The following article is based in part on a presentation made by Kurt R. Hofmeister at the Second Symposium on Quality Function Deployment, held June 18 through 19, 1990, in Novi, Michigan.

ABSTRACT

Quality Function Deployment (QFD) is seeing increased use in the US design/manufacturing arena. This paper examines the role, benefits, and structure of QFD in the service industry. Topics include the House of Quality, modification of the four-stage QFD model for service applications, and guidelines for assessing a QFD study. An example of QFD in a service environment (i.e., a public utility) is also presented.

Kurt R. Hofmeister is a director of the American Supplier Institute (ASI), Allen Park, Michigan.

INTRODUCTION

"The customer lost his temper with the new salesman. But the more he stormed and raved, the more blandly unconcerned the salesman was. 'Doesn't anything I say make the slightest difference to you?' the customer demanded. 'No,' replied the salesman. 'Before I got this job, I was a baseball umpire.' "

The above scenario uses humor to illustrate how a service company—or any type of company, for that matter—*should not* do business. However, situations in which the voice of the customer goes unheard are not uncommon, which is not a laughing matter.

Though most often associated with new product development in the design/manufacturing environment, Quality Function Deployment (QFD) holds equal promise for the service industry, which should also be driven by customer satisfaction. Almost any service organization—from the in-house departments that assist members of other departments to the nationwide department store and restaurant chains that clothe and feed the American populace—can benefit from the QFD process' focus on the customer. That the perception of quality is relevant in the service industry—and the competition for customers fierce—makes the use of QFD even more appropriate.

We can think of QFD as a questioning process used to integrate the voice of the customer into our daily business activities, a process that forces us to ask and answer the question "why?" in relation to customer needs. This questioning process allows us to *understand* our customers and results in a preventive approach to doing business, rather than the reactive approach that often occurs in response to customer *dissatisfaction.* Concurrently, this preventive approach to doing business results in continuous improvement. We get better and better at what we do by *preventing* problems from occurring.

Additionally, the QFD process helps us define how our activities fit into the overall business scheme, which enhances communication among employees and fosters a sense of teamwork and shared responsibility, with all employees working toward the same goals. Thus, not only can QFD improve the quality of the service our company delivers to its customers, it can help improve the quality of our business as a whole.

QFD METHODOLOGY

The House of Quality

The QFD process begins with the collection of data related to the voice of the customer, or the customer requirements. There are many ways to acquire such data, such as market research reports, customer surveys, and focus groups. The purpose is to obtain an intimate understanding of who your customers are, what they want, and why they want it. Asking why allows us to penetrate the surface needs and get to the "root wants." We can then use a simple conversion technique, or even a basic fishbone diagram, to establish the quality characteristics or measurements that our company uses in fulfilling the customer requirements.

We can think of the customer requirements as the *what* items we want to accomplish. We will list these what items and then refine the list to the next level of detail by listing one or more *how* items for each what item (see Figure 6–1). It is important that the how's be objective measurements that are testable and not technical solutions for the customer requirements. We are

FIGURE 6–1
What *Customers Require Is Translated into* How We Measure.

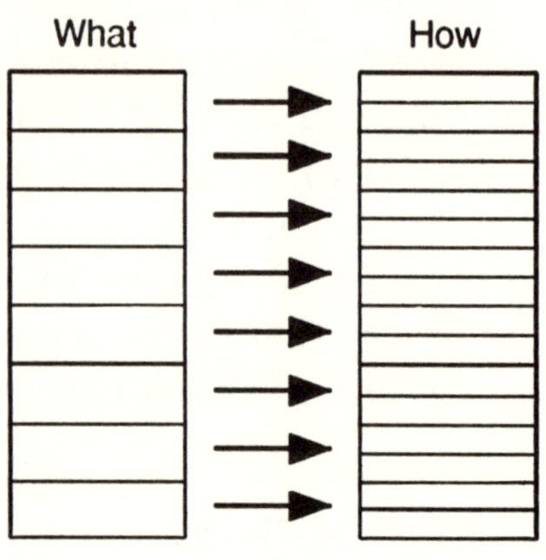

essentially translating the "fuzzy" voice of the customer into measurements in the company's language.

Because some of our how items will undoubtedly affect one or more what items or even adversely affect one another, we will use a matrix chart called the House of Quality to graphically depict the relationships among our what and how items. We will place symbols that depict the strength of these relationships within the House of Quality's relationship matrix (see Figure 6–2). Commonly used symbols are a double circle (strong relationship), a circle (medium relationship), and a triangle (weak relationship).

The House of Quality also contains a section for *how much* items (see Figure 6–3). These are measures for our how items based upon the voice of the customer. How much items serve a twofold purpose: They provide

FIGURE 6–2
Symbols are Used to Depict the Strength of the **What** *and* **How** *Relationships.*

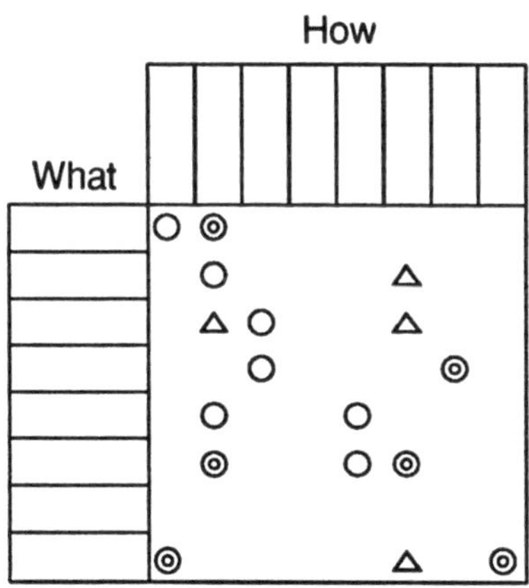

an objective means of ensuring that customer requirements have been met, as well as establishing targets for further development.

We can use several other tables in conjunction with the House of Quality, such as the correlation matrix (see Figure 6–4), the competitive assessment graphs (see Figure 6–5), and the relative importance ratings (see Figure 6–6).

The correlation matrix, a triangular table that attaches to the how items, establishes the correlation between each how item. Symbols similar to those used in the relationship matrix are used in the correlation matrix, such as a circle (positive relationship), double

FIGURE 6–3
How Much *Items Provide Targetable Measures for* Hows.

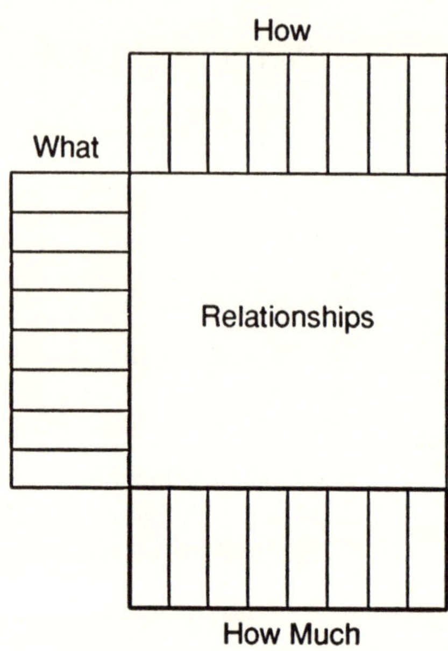

circle (strong positive relationship), X (negative relationship), and double X (strong negative relationship). This is done in order to establish potential conflicts early in the planning process and resolve them through trade-off decision making or developing new service technologies that eliminate the bottleneck.

The competitive assessment graphs depict how competitive services compare with our services both in terms of a customer perspective (against the what items) as well as a technical perspective (how items). This information can assist in straightening out any inconsistency between

FIGURE 6–4
A Correlation Matrix Is Used to Establish Potential Trade-offs.

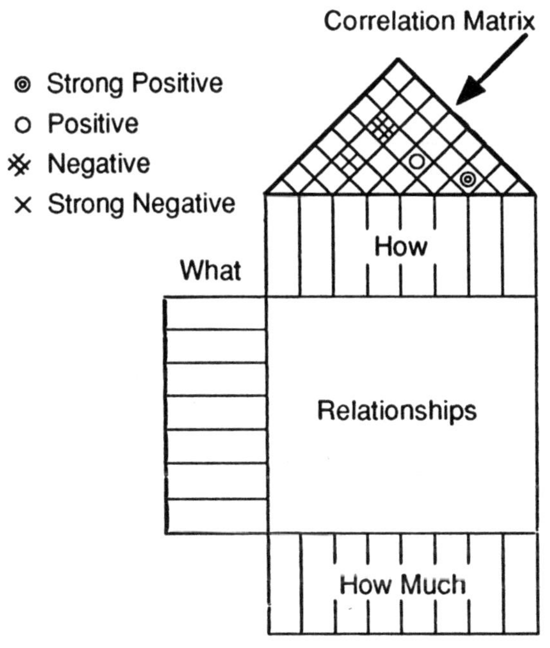

customer perception and technical evaluations and in establishing the company's strategy for improvement (copy a competitor, work on a particular weakness, or innovate where each competitor is performing poorly).

Additionally, we can include numerical tables or graphs that depict the relative importance of each what and how item to the desired end result. The what importance rating is based on customer assessment and expressed as a relative scale, such as 1 to 5. With the how importance rating, weights are typically assigned to the relationship symbols

FIGURE 6–5
Competitive Assessment Graphs Show Our Competitive Position.

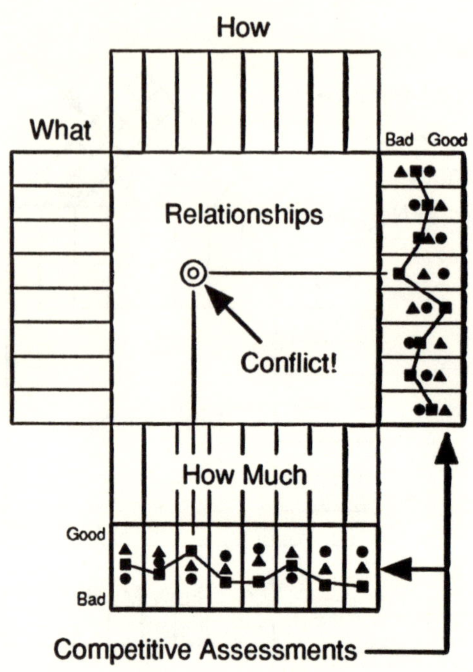

(e.g., double circle equals 9, circle equals 3, triangle equals 1). We then combine the numerical what and how ratings to calculate relationship values for the quality characteristics. Knowing the overall importance ratings will help us prioritize and work on items that yield the greatest level of satisfaction to the customer.

Modifying the QFD Model

The QFD process is often presented as a four-phase model (see Figure 6–7): product planning, parts deploy-

FIGURE 6–6
Importance Ratings Assist in Setting Priorities.

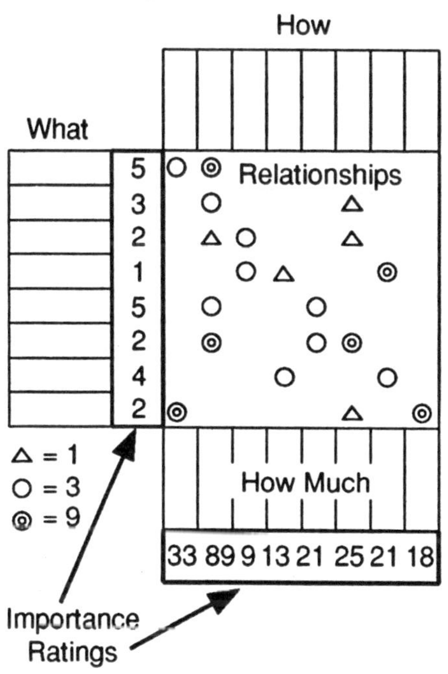

ment (design), process planning, and production planning. The first phase of the process, which involves creating the House of Quality matrix, has seen the most use in US design/manufacturing. We can also modify the four-phase model for use in service-oriented applications. The trick is to identify the best way to use the process in each individual application. For example, we could move directly from phase one (House of Quality) to phase three (process planning) or even have a separate phase three for each of our services.

The American Supplier Institute (ASI), Inc., which conducts both public and in-house seminars and workshops

FIGURE 6–7
The Standard Four-Phase QFD Model for Discrete Manufactured Products.

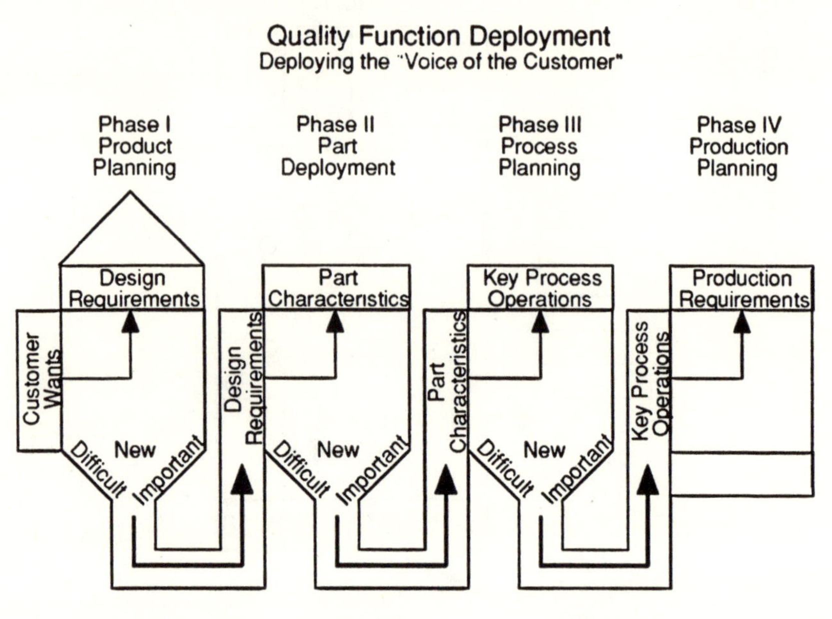

in technologies related to Total Quality Management (TQM), is currently performing a service-related QFD study using a modified model. ASI used a version of the model shown in Figure 6–8 to evaluate its ability to respond to the voice of the customer during seminars and workshops. The resulting service QFD example (see Figure 6–9) provided for service planning (phase one), equipment deployment (phase two), and process deployment (phase three).

What items for the service planning QFD phase include "instructor well prepared" and "can see from back of room." How items for these what items are "marker requirements," "preparation requirements,"

FIGURE 6–8
Modified QFD Model for Service Applications.

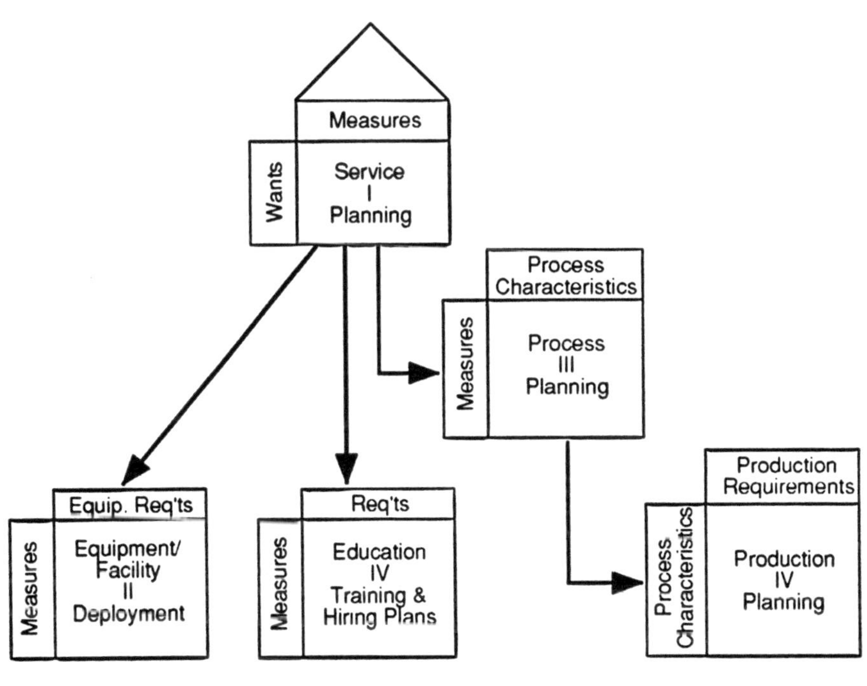

and "text readability." How much items for the how items are "marker readable from 30 feet," "preparation requirements checklist," and "text readable from 30 feet."

Equipment deployment (phase two) and process deployment (phase three) matrices were prepared for each of the shown items, resulting in the introduction of an improved process for instructor preparation and a new brand of marker. Other service industry users of QFD, such as Florida Power & Light Company, an electrical company based in Miami, Florida, report similar process improvements.

FIGURE 6–9
QFD Example for Seminar Quality Improvement.

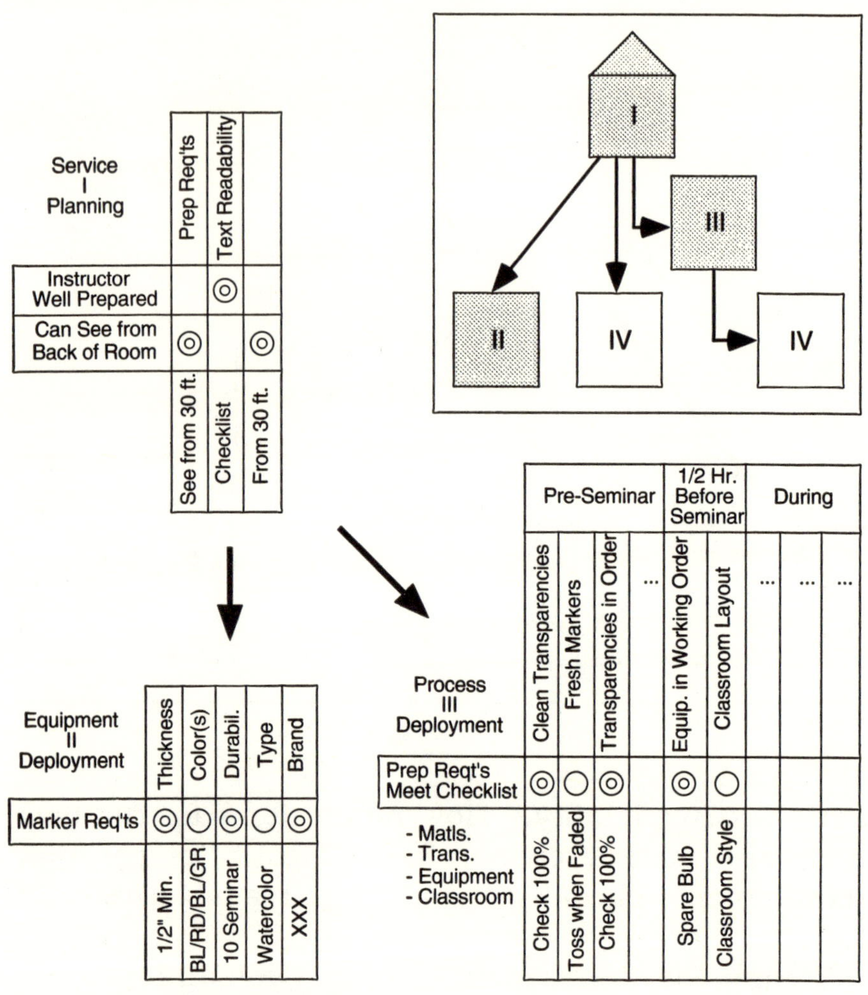

FLORIDA POWER & LIGHT SERVICE APPLICATION

Florida Power & Light (FPL), a division of the FPL Group, uses QFD to identify customer requirements (what items) and quality characteristics (how items) and then deploy them into job functions throughout the organization.

Consultants from Qualtec, Inc., Juno Beach, Florida, also a division of the FPL Group, assisted with FPL's QFD implementation. "Quality Function Deployment at FPL," a paper by Qualtec consultants J L Webb and William C Hayes published in *Transactions from the Second Symposium on Quality Function Deployment*, examines two of FPL's QFD applications: (1) deploying corporate quality elements via a "Customer Needs Table of Tables" and (2) revising the "response to customer calls" process at regional phone centers. The following example is primarily based on the second application (2).

Background

Customers in the area covered by the Southern Division Regional Phone Center, Miami, Florida, generate approximately 40,000 calls per week. FPL employees typically answer five types of calls: requests for payment extensions, connections of new services, questions about billing statements, reports of power outages, or complaints relating to any of the above. Additionally, FPL employees must be able to respond to such calls in three different languages.

Prior to the use of QFD, the following process was in effect:

1. A customer called FPL for assistance.
2. An FPL employee took the call, responded to the customer request, and performed any necessary follow-up.

FPL used a quarterly customer service satisfaction survey to assess customer satisfaction. Problems had been noted in three categories: (1) courteous and professional manner, (2) caring and concern, and (3) ability to answer questions. The Southern Division regional phone center scored low in all three categories.

Creating the Matrices

FPL's QFD team established primary, secondary, and tertiary customer requirements (that is, more and more specific what items) related to the "response to customer calls" process. The primary and secondary what items were drawn from an earlier QFD study; the tertiary items from verbatim responses to open-ended questions on the customer service satisfaction survey, which were then grouped using the affinity process, one of the Seven Management Tools.

The primary what items were "considerate customer service" and "accurate answers/timely actions." The secondary what items for "considerate customer service," for example, were "fair treatment," "courteous, friendly employees," and "concern for customer problems." These items were then further subdivided into tertiary what items. For example, "courteous, friendly employees" broke down into "understandable speech," "professional manners," and "customer courtesy."

The team created the matrix shown in Figure 6–10 in order to examine the relationship between the customer requirements (what items) and the quality characteristics (how items). The customer rating indicated the need to improve the time required to answer phone calls. An action plan and countermeasures were implemented which resulted in a four-fold improvement on performance.

Next, the team analyzed the key quality characteristics in relation to FPL's existing training program and

FIGURE 6–10
FPL Service Planning Example.

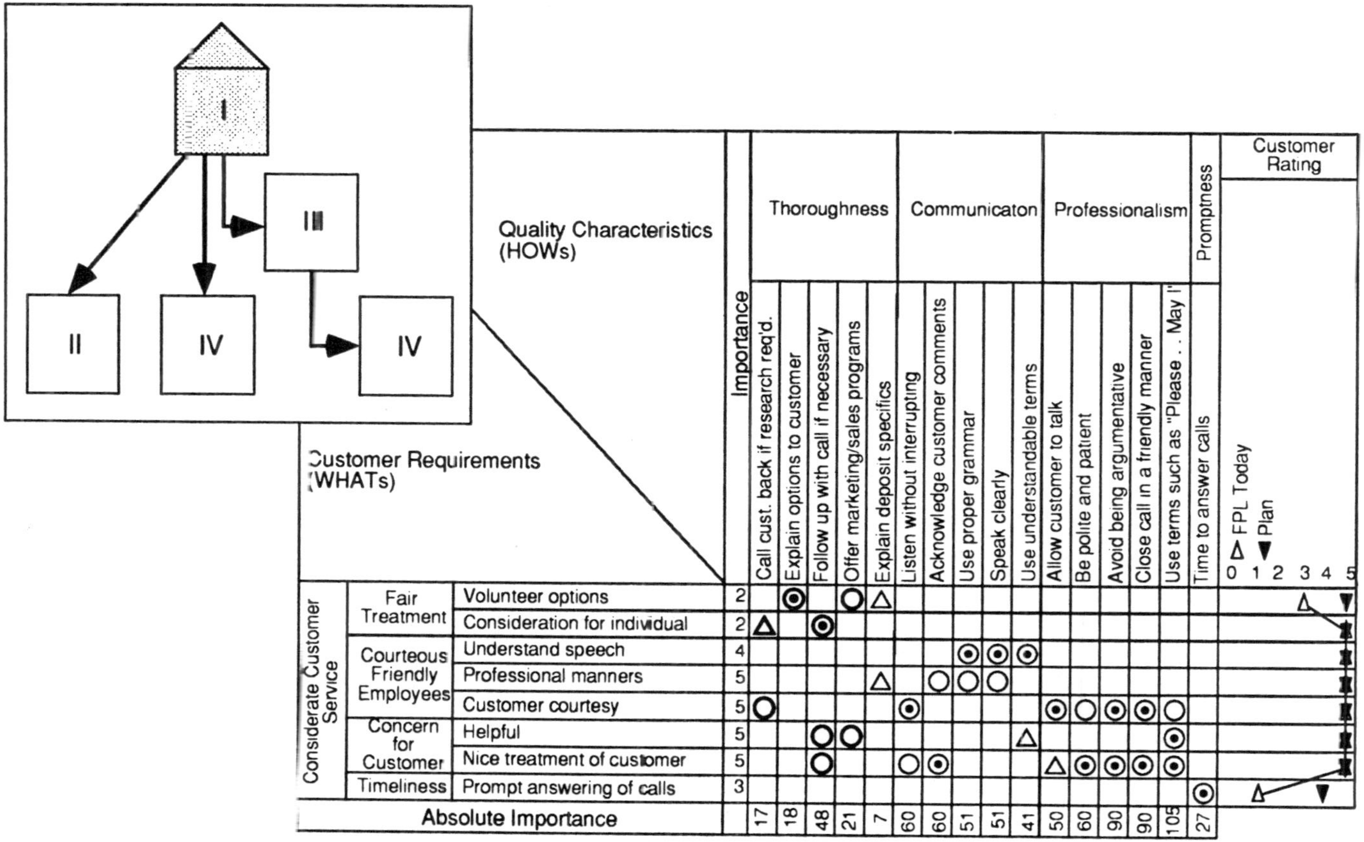

employee selection process and created related phase-four training and hiring matrices (see Figure 6–11 and Figure 6–12). Upon doing so, the team concluded that the key quality characteristics were not being taught in the training program or being considered during the hiring of new employees.

The Revised Process

As a result of the QFD study, FPL made the following changes to how it responds to customer calls:

1. The service observation form used to monitor customer service calls was revised to link the customer requirements with employee performance, providing a more objective evaluation of responses to customers and more timely and valid feedback to FPL employees.

2. The key quality characteristics were incorporated into the existing training program, and customer requirements and job requirements were clearly defined.

3. An interview analysis checksheet was created for use when hiring new employees. The checksheet evaluates each interviewee's ability to perform the required job. The interview itself includes role-playing, which helps the interviewer assess the interviewee's personal skills and reactions and gives the interviewee an idea of the job requirements.

Since the application of QFD to FPL's "response to customer calls" process, customer satisfaction has improved dramatically and customer complaints to the Florida Public Service Commission have been signifi-

FIGURE 6–11
FPL—Education and Training Examples (Phase Four).

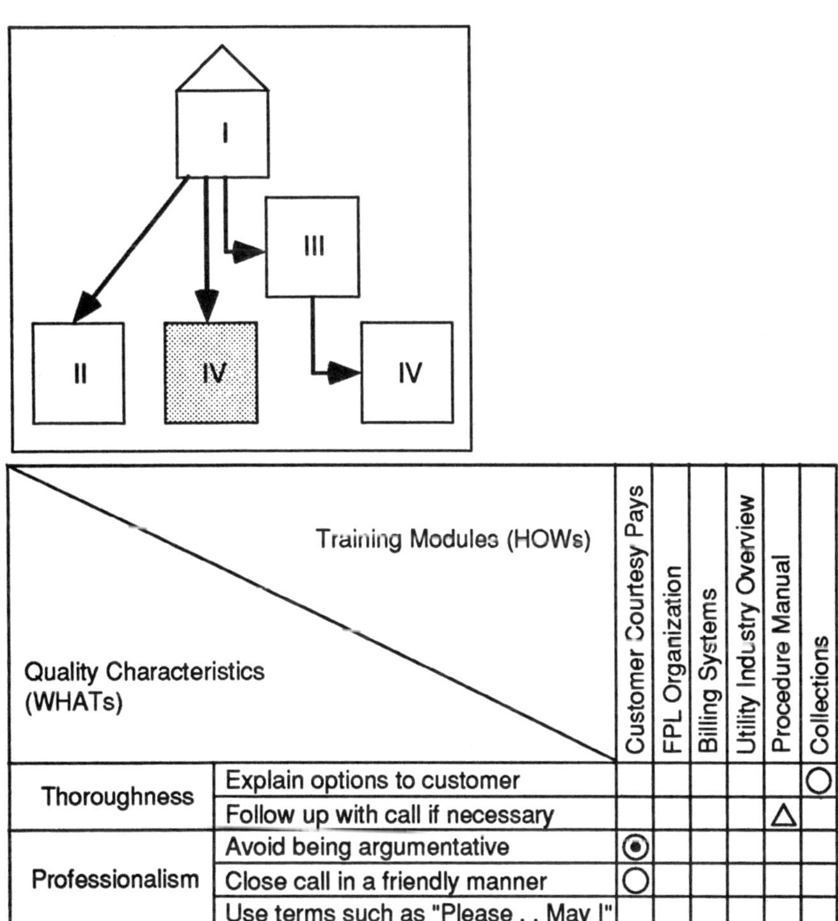

Quality Characteristics (WHATs)		Customer Courtesy Pays	FPL Organization	Billing Systems	Utility Industry Overview	Procedure Manual	Collections
Thoroughness	Explain options to customer						O
	Follow up with call if necessary					△	
Professionalism	Avoid being argumentative	⊙					
	Close call in a friendly manner	O					
	Use terms such as "Please . . May I"						

cantly reduced. Additionally, regional phone center management has become more aware of how different activities affect customer satisfaction; specific problems have been passed on to quality improvement teams; and

FIGURE 6–12
FPL Hiring Matrix Example (Phase Four).

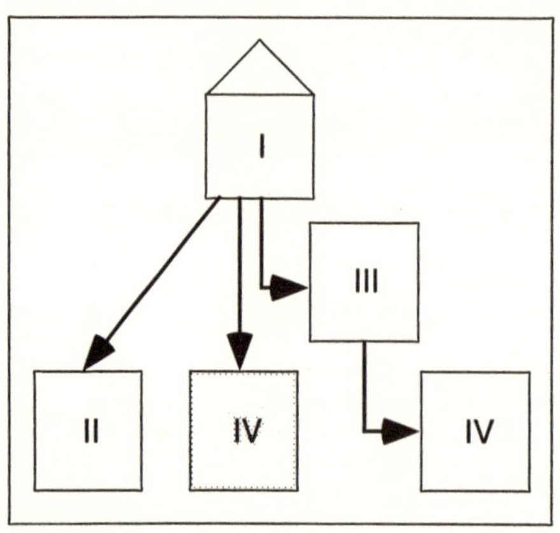

Quality Characteristics (WHATs)	Hiring Elements (HOWs)	High School	College	Military	Work Experience	Attitude	Self Assessment	Communication Skills	Appearance
Thoroughness	Explain options to customer								
Thoroughness	Follow up with call if necessary								
Professionalism	Avoid being argumentative					△	△	○	
Professionalism	Close call in a friendly manner							○	
Professionalism	Use terms such as "Please . . May I"							○	

additional QFD studies have been initiated in service, software, and product development areas.

GETTING STARTED WITH QFD

Managing the QFD Process

As such users as FPL can attest, in order to realize the full benefit of QFD, the voice of the customer must be integrated into your everyday activities. If QFD is not tied into the existing systems in place at your company, it will die a quick death. And in order to be tied into the existing systems in place at your company, QFD needs management commitment and endorsement, as well as personal commitment from each QFD team member.

The first step in a QFD study is identification of the project and the team members. The first project should be fairly simple and small in scope, and the people selected to work on it should come from a cross section of job functions and/or departments and have the expertise and the desire to participate in the study.

The second step is to seek training and/or outside consultation. Many companies find outside facilitators very helpful when beginning with QFD. At the very least, team members should receive adequate training in QFD methodology from a reputable source.

Once trained, team members must be given enough time to successfully perform their responsibilities. At each team meeting, members will coordinate activities, review and update charts, analyze information, and decide on what additional information needs to be gathered. Most of the real work, however, occurs outside the team meetings, with members gathering information for use at the next meeting.

ASSESSING YOUR EFFORTS

It is essential that you assess your QFD implementation on a regular basis. The following guidelines, which summarize much of what has been previously stated, will give you an idea of what to strive for and, conversely, what to avoid.

Less is more. Strive for a few well managed, focused QFD projects versus numerous projects accompanied by little thought or supervision.

Customer satisfaction first. The objective is to delight the customer, not to make QFD charts.

Plan now, act later. Establish your purpose, scope, goals, objectives and strategy before beginning a QFD project.

Keep it simple. Choose your first project carefully, opting for small and manageable versus large and complex.

Prevention is the key. Use QFD to prevent problems (proactive), not to put out fires (reactive).

The front line. Line management should do the QFD project, not the staff.

By invitation only. Create an integrated, cross-functional team of experts for each QFD project. Do not settle for whoever is available. However, team members should be willing volunteers, not draftees.

A true team. Team members should operate with a common purpose and goals, not as individuals in one room defending their own turf. Team members should say "What's in it for us?" not "What's in it for me?"

A true believer. The team leader should be strong and positive—a true believer—not a cynic of QFD.

Just-in-time training. Team members should be trained together after the QFD project has been identified, not "just in case" a project begins.

Facts, not numbers. Management should know about the product and process improvements, not just how many QFD charts there are.

No 24-hour days. Team members should be given time to prevent problems, not be kept busy fighting fires. QFD should be integrated into the everyday activities.

A true leader. Management should participate and lead the QFD project, not merely be involved.

Do not be alarmed or discouraged if your initial experience with QFD is not 100 percent positive: Virtually every successful QFD study experiences difficulty at one point or another. Like almost anything worth learning, QFD takes time and effort. Experience is the best teacher.

ACKNOWLEDGMENT

I would like to thank Nancy E. Ryan for her assistance in the writing of this article.

REFERENCES

American Supplier Institute, Inc. *Quality Function Deployment Awareness Manual.* Dearborn, MI, 1989.

Braud, Jacob M. *New Treasury of Stories for Every Speaking and Writing Occasion.* Englewood Cliffs, NJ: Prentice-Hall, Inc. 1959.

Eureka, William E. and Nancy E. Ryan. *The Customer-Driven Company: Managerial Perspectives on QFD.* Dearborn, MI: ASI Press, 1988.

Hayes, William C, and J L Webb. "Quality Function Deployment at FPL," *Transactions from the Second Symposium on Quality Function Deployment.* Planning Committee for the Second Symposium on Quality Function Deployment, sponsored by the American Supplier Institute, Inc., Dearborn, MI; the Automotive Division of the American Society for Quality Control, Northville, MI; and the Growth Opportunity Alliance of Lawrence/Quality-Productivity-Competitiveness, Methuen, MA, 1990.

Hofmeister, Kurt R. "Applying QFD in Various Industries," *Transactions from the Second Symposium on Quality Function Deployment.* Dearborn, MI, 1990.

Chapter Seven

Customizing Your House

William E. Eureka

Nancy E. Ryan

DIFFERENT STYLES OF HOUSES

After the customer requirements have been translated into company jargon, that is, company measures or engineering terminology, the first House of Quality can be built. But what if there are too many customer requirements and company measures to count? What if there are scores of these requirements, say 100 or so?

This many requirements cannot be managed in a traditional House of Quality. Luckily, however, other housing options exist.

Condos of Quality—the subsystem-level of Quality Function Deployments that help keep the QFD project

This chapter is excerpted from the second edition of *The Customer-Driven Company: Managerial Perspectives on QFD*, Irwin Professional Publishing and the ASI Press, 1994. William E. Eureka is a learning coach at Herman Miller Inc., Zeeland, Michigan. Nancy E. Ryan is a free-lance writer in Oakland, Michigan.

efficient and focused—can be used with a large number of customer voices. Managing a large number of voices can seem nearly impossible with only one matrix. The time requirements for determining the relationships and obtaining the supportive data are often overwhelming in such situations, not to mention discouraging.

Condos of Quality are better housing options when the number of voices exceeds 25, although up to 50 voices can be manageable with the traditional House of Quality, depending on the team and the project.

A preplanning matrix is usually used initially to prioritize and identify which issues to attack. This often leads to subsystem QFD studies (i.e., Condos of Quality). To put this another way, if the product is as simple as a lawn sprinkler, a detailed QFD study could be done on it. But if the product is more complex, it is better to use a subsystem approach and the preplanning matrix, with the following exceptions: (1) when the product is so simple that a preplanning matrix is not required and (2) when the priority decisions that the matrix would make have already been made by other means.

After the preplanning matrix has been carefully completed, work on the subsystem QFDs can begin (see Figure 7–1). The QFD process for the House of Quality applies to the Condo of Quality as well, with one addition: An interaction matrix (see Figure 7–2) documents any system-to-system correlations.

LIAISONS AND LINKAGES

QFD has perhaps the greatest potential for success when linked with a company's total business system; this includes marketing. Collaborating with marketing, how-

FIGURE 7–1
Complex Products Call for Subsystem QFD Studies,
That Is, Condos of Quality instead of Houses.

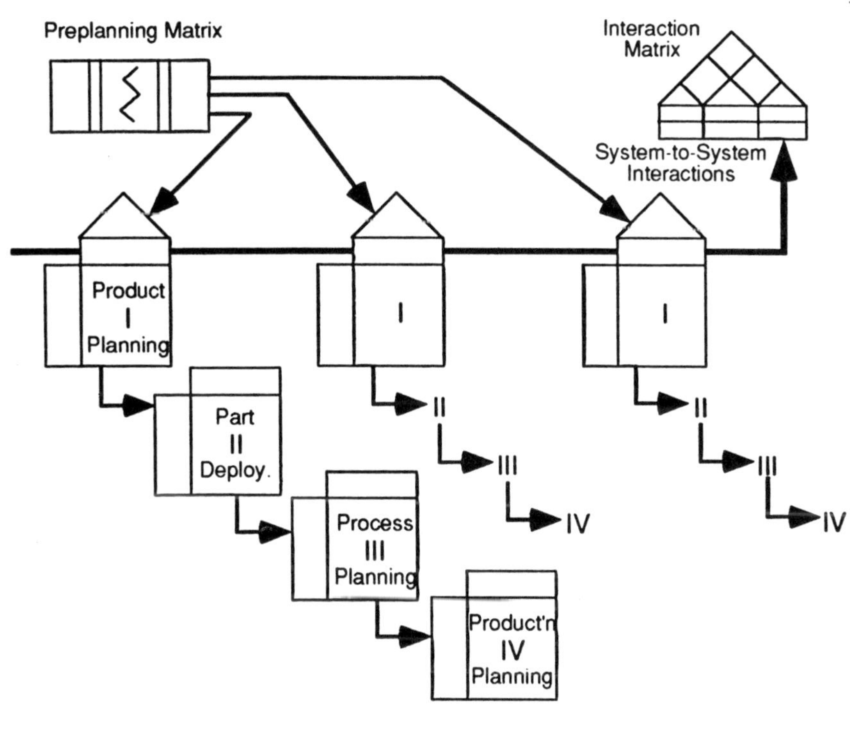

ever, can sometimes be problematic because the marketing profession has an established tradition of how to do market research, and that tradition does not include QFD.

How does the market research that precedes the first stage of QFD differ from traditional marketing? To begin with, QFD is more of a qualitative approach, whereas traditional marketing is more of a quantitative. And although marketing people study the same problems,

FIGURE 7–2
The Interaction Matrix Documents System-to-System Correlations That Occur between Subsystem QFD Studies.

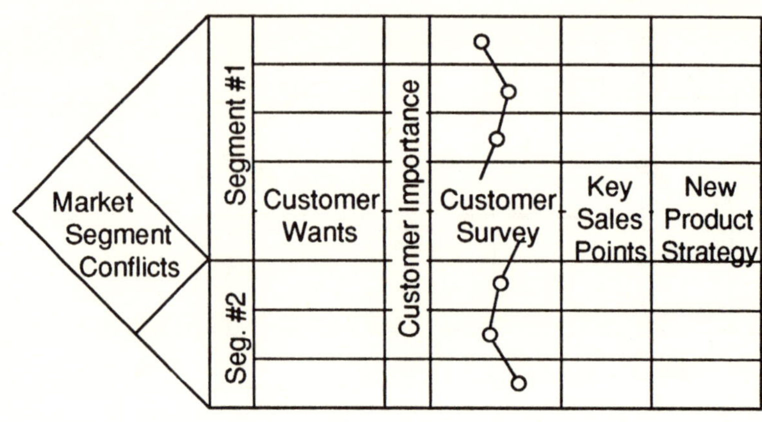

QFD has evolved from a completely different body of knowledge and may initially be viewed as a suspect activity. But despite some basic differences, both QFD and traditional marketing research are working toward a common goal: a clear understanding of the voice of the customer. Thus, marketing people that see QFD in action typically buy into it. The proof is in the House of Quality.

Kevyn Irving, manager, technical services, contract development, used a "skeleton in the closet" approach to sell QFD to the marketing department and board of directors at Ethicon Endo-Surgery, a division of Johnson & Johnson based in Cincinnati, Ohio. Due to the rapidly changing market demands in the booming field of endoscopic medical devices, time to market is a prime factor in Ethicon Endo-Surgery's business strategy.

In the traditional medical device field, a company can rapidly design a new product, move it into the market-place, and drive the costs out while absorbing the losses. In the endoscopic medical device market, however, the product might very well be obsolete before the costs are driven out. Thus, new products are typically rushed to market in response to marketing department research but at significant expense. This expense can be devasta-ting if the products fail.

As Irving noted, "People want to see results right away versus taking the time to sit back and really see what the customer wants. The problem is, when you rush right out like that, you're going to get it wrong far more than get it right.

"In my opinion, QFD really flourishes here," Irving added before describing a QFD trial that could have saved the company months of development, retooling time, and significant funding. Interestingly enough, the trial QFD study identified 32 product features that were not in the revised product description, the majority of which are now considered important customer require-ments, such as interchangeable shafts with attached end effectors and two different handle designs.

"This study added more credibility to our use of QFD," Irving noted. "There's a QFD chart in my office right now that's on its way to marketing. QFD's really getting into the system here." In addition, Ethicon Endo-Surgery is moving toward organizational changes that will en-hance the team concept, an important aspect of QFD.

A technique coined Matrix Data Analysis can be employed during the marketing research stage of QFD. This tool, which represents the correlations found on the matrix chart on an x and y axis, is used to geographically portray market segmentation and to identify appropriate markets for both the user's company and the competition.

Although QFD can also be used for strategic and business planning, Policy Management (Policy Deployment) is the better choice. Policy Management is the process for developing achievable business plans and then deploying them throughout the company. It closely resembles QFD, which is a better choice. The *what/how/ how much* process used during a QFD study is nearly identical to that used to chart the goals and actions in Policy Management. The process results in a series of cascading matrices that closely resembles the House of Quality. (See *The Process-Driven Business: Managerial Perspectives on Policy Management,* ASI Press, 1990, for more on this subject.)

While applying QFD to strategic or business planning will result in the development of individual plans and actions to accomplish corporate, divisional, and departmental strategies, Policy Management keeps a few extra items in its toolbox. These include the Plan-Do-Check Act (PDCA) Cycle and "catchball."

QFD and Policy Management are definitely complementary, however, and should be used in conjunction with one another whenever and wherever possible. First-tier Japanese companies, for example, often use Policy Management to solve their business planning needs and then apply QFD to carry their products through the product-development process. Several American companies, including Florida Power & Light Company and The Budd Co., Troy, Michigan, have also used QFD in conjunction with Policy Management.

To quote Norman E. Morrell (1988), corporate manager, quality product reliability at Budd, "Just as QFD translates the wants of the customer into actions that create a product, Policy Management ensures that your planning strategies are customer-sensitive" (*The Process-Driven Business* p. 73).

The New QFD

The Fifth Symposium on Quality Function Deployment featured 45 case studies related to QFD, a dozen of which were pertinent to service-related and "soft" issues. The case studies covered a wide range, from "Applying QFD to Health Care Services: A Case Study at the University of Michigan Medical Center" and "The Application of QFD to the Los Angeles River Rescue Task Force" to "QFD and Personality Type: The Key to Team Energy and Effectiveness" and "QFD in Academia: Addressing Customer Requirements in the Design of Engineering Curricula."

The abstract for "Applying QFD to Health Care Services: A Case Study at the University of Michigan Medical Center," by Dr Deborah M. Ehrlich and Dennis J. Hertz, for example, reads as follows:

> The University of Michigan Medical Center (UMMC) piloted Quality Function Deployment (QFD) in a new unit that consolidated several separated diagnostic procedures into one unit. Based upon early TQM successes, the organization employed QFD to realign resources to meet the valid customer requirements of the combined groups in order to stimulate service volume by better satisfying customer desires. This paper will discuss the UMMC QFD approach, articulate experiences learned, identify changes that have been implemented, quantify the financial benefits that have resulted from these changes, and offer ideas on how to best utilize QFD at a referral hospital.

As a result of the increased interest in QFD for use in such diverse settings as hospitals and academia, both

ASI and GOAL/QPC are offering increased training and course work on QFD for the service and administrative sector. The Quality Service Model featured in ASI's *QFD for Service Implementation Manual* is based on three key tracks: (1) plan for success, (2) improve the process, and (3) hold the gains. This conceptual model incorporates such QFD-dictated tools as the Kano model, preplanning matrix, House of Quality, Pugh Concept Selection, process flow charts, and control plans. The conceptual model consists of the following steps:

1. Plan for success
   - Plan the process.
   - Make the process visible.
   - Develop the preplanning matrix.
   - Develop a detailed House of Quality on key opportunities.
2. Improve the process
   - Develop new processes/alternatives.
   - Apply Pugh Concept Selection.
   - Try out new processes/equipment.
3. Hold the gains
   - Document critical process steps.
   - Standardize the process.
   - Publicize the results.

ASI has also adapted the QFD four-phase model for use with continuous/batch processes.

QFD AT WORK

The following real-life examples illustrate how novel the application of QFD can be. The company names have been changed in respect of proprietary information.

FIGURE 7–3
J & N Corporation Sharpened Its Product-Improvement Effort via These Customer Requirements, an Importance Rating Based on Customer Input, and a Customer Competitive Assessment.

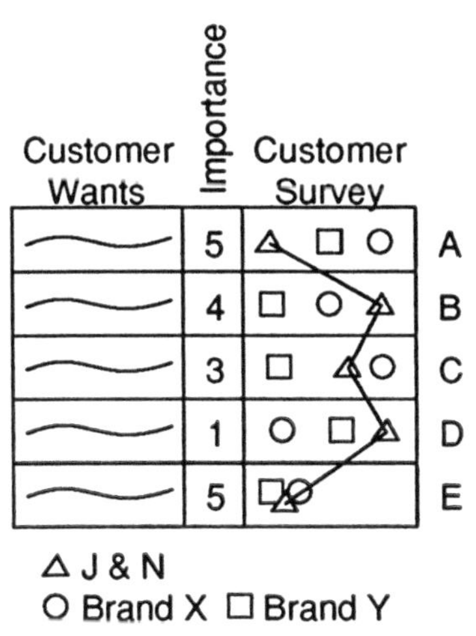

J&N Corp.

J&N Corp., a large consumer goods manufacturer, had access to more market research than it knew what to do with. Sifting through this market research and trying to glean something meaningful from it was a laborious and often subjective task. QFD provided a way to organize the data so that J&N could effectively use it.

By using the preplanning matrix shown in Figure 7–3, J&N sharpened its product-improvement efforts. It used the customer requirements, an importance rating based on customer input, and a customer competitive assessment.

Although the first customer want (A) was very high in importance, the competitive assessment showed that J&N was trailing the competition in this area. Thus, it would probably want to make an improvement.

The second customer want (B) was also high in importance, but J&N was the leader here, with the key competitors doing the trailing. Hence, the company decided to do nothing except maintain its lead.

The third customer want (C) had middle-of-the-road importance. Since J&N was middle-of-the-road in terms of competition, no action was deemed necessary.

The fourth customer want (D), which J&N was also leading, wasn't that important to the customer. Although J&N didn't wish to hurt its position here, it was willing to compromise in order to make a strong improvement elsewhere.

The fifth customer want (E) is an important requirement that neither J&N nor its leading competitors had satisfactorily addressed. Thus, this want presented a real opportunity for J&N to leave the competition behind.

B&B Steel

B&B Steel, a major US steel company, wanted to build a new steel mill. Because steel mills are designed so infrequently and are so expensive to build, not to mention very difficult to change once they have been built, forethought was of utmost importance.

B&B selected QFD as the best tool to solve this problem. Its team created a QFD chart (see Figure 7–4) in which the what items were the customer wants for the steel to be produced (which covered a wide range, since steel mills make many types of steel). The how items

FIGURE 7–4
***B&B Steel Used QFD to Foolproof the Design of Its
New Steel Mill.***

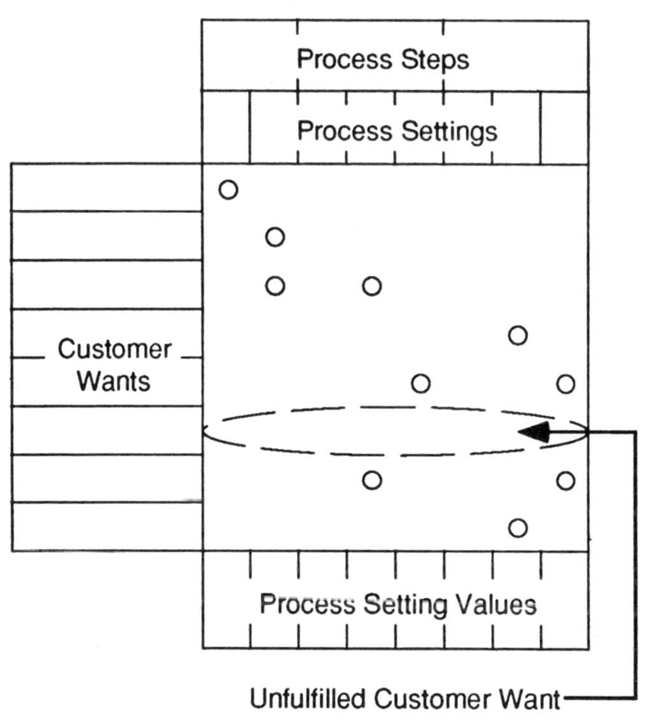

were the process steps and the process settings for each
of the process steps, and the how much items were the
setting values.

After completing the chart, the B&B QFD team noticed
some blank rows, which equated to unfulfilled customer
wants. This led to a number of improvements in the
company's new mill design, improvements that would
have been extremely difficult or downright impossible
to achieve after the mill had been built.

Baker Electric Co.

The Baker Electric Co., a small but prospering electric power utility, had an unusual problem. Whereas most large companies can wield a big stick over their suppliers, most of Baker Electric's suppliers were larger than Baker Electric itself. The primary supplier from which Baker Electric purchased its transformers, for example, was huge—so huge that it considered Baker Electric ''small.''

So Baker Electric decided to use QFD for itself, identifying the customer requirements for a substation transformer (see Figure 7–5). These requirements became the what items and the actual transformer specifications became the how items. Based on this data, Baker Electric put together a comprehensive set of specifications that it then gave to the supplier, thus ensuring that its voice be heard.

Ashton Corporation

Ashton Corporation, a large mechanical device manufacturing company, established a cost-reduction program that included reducing the costs of some of its parts. These parts tended to be very complex and have a multitude of specifications. Because the parts were considered critical, most of the specifications automatically had tight tolerances.

The QFD chart shown in Figure 7–6 provided a closer look at the parts and their specifications. Ashton Corp.'s QFD team completed a chart for each part with cost-reduction potential and listed the customer wants for that particular part. These included what the part was supposed to do for the customer, that is, the what items.

FIGURE 7–5
***Baker Electric Company's QFD Study Yielded a Set
of Specifications That It Passed Along to Its Much
Larger Supplier.***

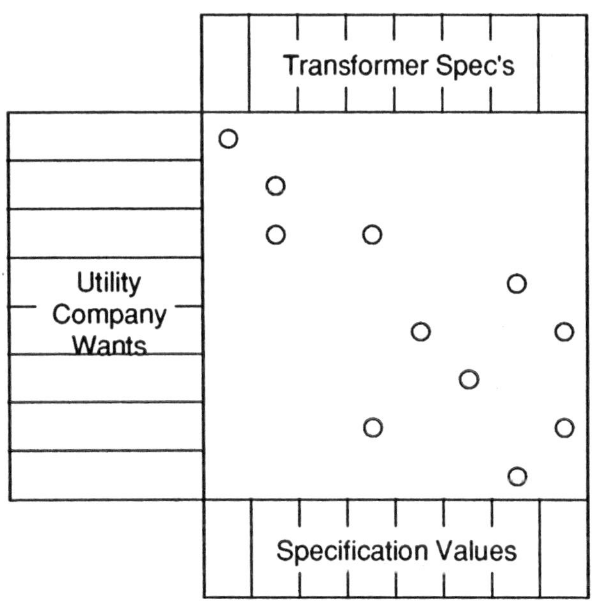

The how items were the specifications that came right
off the part's blueprint.

The resulting importance ratings for each of the how
items gave an indication of the relative importance of
that item to the customer. In many cases, the columns
came up blank—the specifications had nothing to do
with satisfying the customer wants.

This actually wasn't an error; many of the specifications simply had to be there in order to manufacture
the part. Because these specifications weren't crucial
to satisfying the customer wants, their tolerances were

FIGURE 7–6
***Ashton Corporation's Cost-Reduction Program Bene-
fited from QFD Analyses of Critical and Noncritical
Parts.***

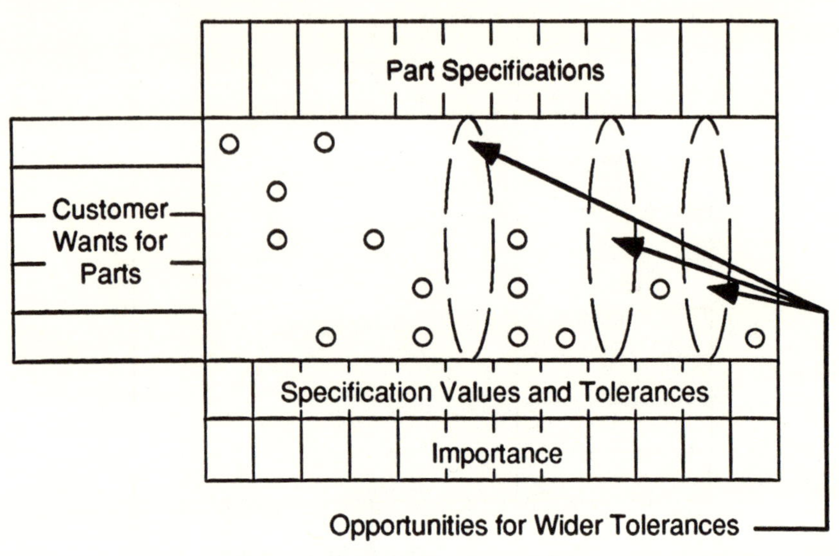

opened up, which resulted in significant cost savings.
Specifications with many symbols, however, were considered critical and their tolerances were maintained.

Saranen Enterprises

Saranen Enterprises wanted to participate in the European Community market when it opened up in 1992,
but needed a European partner. Being a small company
and having heard a number of horror stories about joint
ventures, it set out to systematically find the *right*
partner.

FIGURE 7–7
Saranen Enterprises Applied QFD to the Identification of a Joint-Venture Partner with Which It Could Participate in the European Community Market.

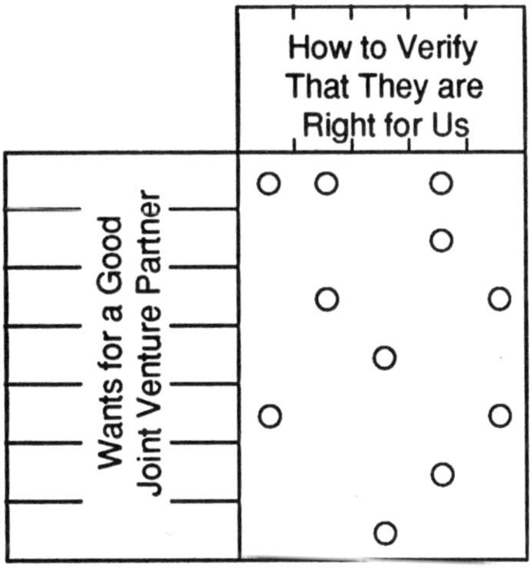

By using QFD, Saranen Enterprises identified its expectations of a good joint-venture partner (see Figure 7–7). The how items identified whether a potential joint-venture partner met those expectations. These were specific questions that Saranen Enterprises could ask and verify instead of relying on the spoken word and trust.

Through this process, Saranen Enterprises identified three prospective joint-venture partners and then narrowed the three to one. Within two days of that decision, it had negotiated a successful joint-venture arrangement. Moreover, the joint-venture partner was astounded that Saranen Enterprises had asked the right questions, questions that the partner hadn't considered asking.

These examples are based on actual uses of QFD by American companies. A more detailed example of how one leading US utility applied QFD to its "response to customer calls" process can be found in the preceding chapter ("QFD in the Service Environment" by Kurt R. Hofmeister).

A BLUEPRINT FOR YOUR QFD

A comprehensive description of how to do QFD is beyond the scope of this book. The following eight steps, however, provide a basic roadmap to QFD implementation. Definitions for some of the technical terms are found in the glossary.

Step 1: Choose the Project Scope

A core team of QFD participants should be selected to define and deploy the project scope. Many first-time QFD efforts are simply too big and overwhelming. The fundamental assumption behind QFD is an important one. The average company's normal operating system works well enough most of the time or that company wouldn't be in business. A key question to ask is, What do we expect QFD to accomplish that our normal product-development process would not? If no good answer exists, the project scope probably isn't correct. Several aspects of the product-development process are typically risky and do require special attention from QFD.

Step 2: Identify Customers and Their Wants

Every company has a full series of customers (multiple customers) no matter how simple its product or market. These customers can be viewed as a value chain extending from the company itself to its various distributors and customers, the end users, and, in more and more instances, product disposal. (This is particularly true in Europe, where disposal of excess packaging materials is increasingly difficult, and in the medical-products field due to biohazard considerations.)

Customers come in two varieties: those the normal product-development process can handle and those with special needs. In addition, some of the latter have high-risk special needs. Once identified, these needs (customer wants) will typically flow into the House of Quality. At this point, the Kano model can be used to explore basic performance and excitement features and achieve maximum competitive advantage.

Step 3: Determine the Technical Requirements

The customer wants identified in step two can now be translated into technical requirements, an internal engineering restatement of the wants. These become the specifications for the overall product. The what items (customer wants) from the House of Quality will be mapped into how items (technical requirements), relationships will be defined between the what and how items, and values will be determined (the how much items).

Determining the how much values (outputs of the QFD chart) often requires additional rooms in the House of Quality. These additions include the rooftop (to look at technical conflicts between specifications), competitive-assessment graphs to depict the state of the marketplace, and importance ratings to assist prioritization.

Step 4: Create a Design Concept

The first phase of the House of Quality should now be complete. The next phase, design deployment, identifies the best design concept, the one that satisfies the technical requirements at the lowest possible cost. Basic creativity fueled by competitive benchmarking and tools like Value Analysis/Value Engineering and Design for Assembly are useful at this point.

A variety of concept alternatives, the seeds of the ultimate design, will result from the above efforts. The challenge is to determine which is the best concept. Pugh Concept Selection can be applied here, with technical requirements, cost, and important business issues all influencing the selection. This approach contrasts the possible concepts, resulting in a synthesis of new, improved ideas and, ultimately, the best design.

Step 5: Choose Part and Material Values

The concept can now be detailed out, right down to the bill of materials in which parts and materials are defined. The part characteristics and material qualities that make the product design achieve the technical requirements are placed in the top of the design deployment matrix (the how items). A process FMEA can be used to identify which of the many part characteristics are particularly critical. Taguchi Methods can then be applied to help identify the values for the part characteristics (the how much items). This activity concludes design deployment, the second QFD phase.

Step 6: Select the Process

Step six begins the third QFD phase, process planning. Here, the process steps used to create the fabricated parts and assemble them are identified. Existing processes will typically be used so as to avoid the capital investment that accompanies new equipment. Although new equipment should be purchased when capacity or capability dictates, cost implications and process capability must be carefully considered. The individual process steps can than be mapped out.

Step 7: Choose the Process Parameter Values

Next, the critical process parameters for each process step—for example, those knobs and dials that adjust to make the process operate effectively—are identified. The parameter values can then be determined. Less critical values will be based on current production practices; more critical ones are determined by Taguchi Methods. This concludes the third QFD phase.

Step 8: Choose the Control Methods

The last step concerns the fourth QFD phase four—controlling manufacturing operations. Types of human intervention to help control the process on a daily basis can now be identified, just as the settings for the process parameters were determined in step seven. Step eight addresses practices and procedures rather than engineering issues. These are based on a risk assessment driven by a process FMEA or, if not available, the QFD team's best judgment. The control methods will fall into four categories: quality assurance, preventive maintenance, mistake-proofing, and standard operating practices that ensure consistency in the operation.

These eight macro steps define a roadmap for implementation of QFD and a means of tracing critical part characteristics through the manufacturing process. Organizations in the aerospace industry that must comply with Boeing's D1–9000 requirement will find a natural means of ensuring traceability in QFD. Organizations in other industries will benefit similarly.

Chapter Eight

"I Can't Get No Satisfaction"

Jim Quinlan

Diane Byrne

Today, many successful companies have come to realize that increased profits result from business practices directed toward satisfying the customer. In fact, many enlightened executives are now preaching that competitive leadership necessitates not merely a company's ability to *satisfy* its customers, but also its ability to *excite* customers by exceeding their expectations. Meeting and exceeding customer requirements while shortening product development cycles and reducing costs are key objectives for companies that are serious about competing in today's global marketplace. The critical question is, how can these objectives be achieved?

The task of developing products to meet and exceed customer requirements includes two elements: (1) making sure the product contains the right features and (2) optimizing the product's performance. Clearly,

Diane Byrne and Jim Quinlin are principals of ITEQ International, a consulting firm in Livonia, Michigan.

ensuring one of these elements without the other is not sufficient. For example, it is not enough to make sure a particular automobile has an automatic transmission. The automobile company must also make sure the transmission smoothly changes gears in a timely manner. In other words, the transmission must exhibit good performance. Conversely, if the customers for a particular make of automobile prefer a manual transmission, then even a car with an automatic transmission exhibiting the highest performance imaginable will probably not sell well. Companies must ensure both features and function.

To ensure that a product contains all the right features as far as the customer is concerned, an approach called Quality Function Deployment (QFD) can be used. QFD is a disciplined planning method that focuses on understanding what the customer wants and needs and then translating these customer requirements throughout product development and manufacturing to make sure the resulting product meets or exceeds customer expectations. The four phases of QFD are characterized in Figure 8–1. The effectiveness of QFD implementation depends on how well a company can work crossfunctionally to understand what the customer wants, to establish priorities, and to maintain the linkage from one phase to the next.

After the customer wants have been identified in the early phases of the QFD process, the team of managers and engineers must translate the wants into measurable requirements. They must then select and optimize a design to achieve high performance at the lowest possible cost. Quality Engineering, as developed by Dr Genichi Taguchi (also called Taguchi Methods™), is a method of engineering optimization that cost-effectively

FIGURE 8–1
Quality Function Deployment—Deploying the "Voice of the Customer."

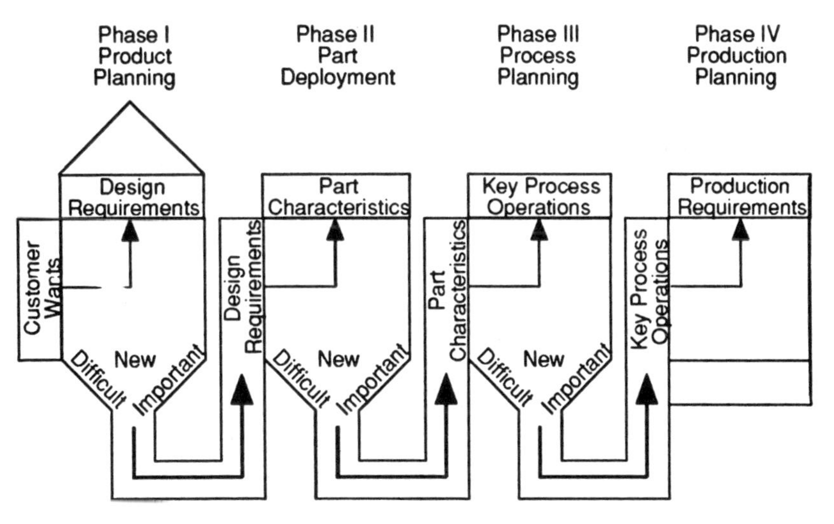

improves the performance of products and manufacturing processes. The further upstream these methods are applied, the greater the benefit.

From the perspective of an entire company, the greatest benefits are often realized when Taguchi Methods are applied in technology development. Taguchi Methods are employed throughout product/process planning and development (the first three phases of QFD) to optimize product and process performance and as a means to evaluate different product concepts and manufacturing processes.

Through the combined use of QFD and Taguchi Methods, a company can develop a product that contains the right features and exhibits high performance as evaluated by the customer, while at the same time realizing significant cost savings and shorter development cycles.

Figure 8–2 shows the major steps that should be undertaken when developing a product that will meet or exceed customer requirements. This paper discusses how QFD and Taguchi Methods can be employed, particularly in the early stages of product planning and development, to ensure that these activities are efficient and effective. We'll begin with understanding what the customer wants from the product.

WHAT THE CUSTOMER WANTS

One of the most difficult tasks for business executives is to find out what the users of their products—their customers—really want. Many times, US companies begin their product development efforts with a fuzzy understanding of what is important to customers, how the product is to be used, and how their product compares to their competitors'. Yet the answers to these questions must be the drivers for product development if the resulting product is to be well accepted in the marketplace.

The most important source for determining customer wants and needs is, of course, the customer. Managers should approach customers directly to find out what they want. They should begin with the customers' own words. This is often called obtaining the "voice of the customer" and can be accomplished through interviews, market surveys, and a variety of other methods.

In addition, a company can discover a great deal about customer needs by observing customers as they use the product. This indirect approach often catalyzes the cre-

FIGURE 8–2
New Product Development.

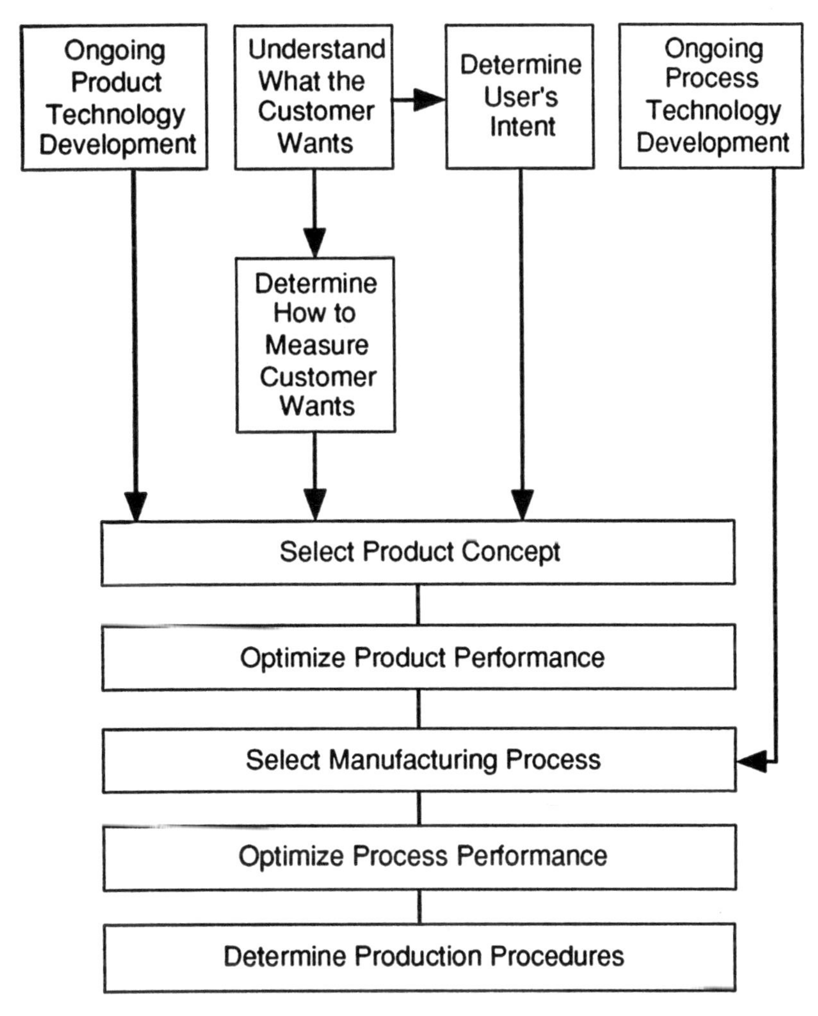

ative thinking of managers and engineers when, for example, they see a customer become frustrated with a particular feature or lack of feature. To determine customer requirements, it is generally best both to watch and to listen to the customer.

Translating the Voice of the Customer

Once customer requirements have been determined, they should be reviewed for clarity. It is sometimes necessary to translate the original requirements into simple, nonrestrictive statements of a single meaning.

For example, Table 8–1 contains a partial listing of customer wants for a toy water pistol, as voiced by a group of children ages 6 to 13. One child vocalized the requirement "easy-to-pull trigger." This customer requirement, as expressed, assumes that the product will necessarily contain a trigger. The likely intent of the child's statement is that the water pistol be "easy to shoot." (The child probably does not care too much about whether or not the water pistol has a trigger, but merely assumed it would since most do.) So as not to be restricted to any specific product concept at this preliminary stage, the engineer translated the original voice of the customer, "easy-to-pull trigger," into "easy to shoot." Similarly, the customer requirement "trigger doesn't pinch finger" was translated into "doesn't hurt hand."

Engineers should avoid the temptation to take the raw, customer-expressed requirement and immediately translate it into a more specific design requirement. For example, the water pistol's engineer may have been immediately inclined to translate "easy-to-pull trigger" into "force requirement to displace trigger 10 mm." This practice of anxious engineering can inhibit innovation and should be resisted at the preconcept selection stage.

TABLE 8–1
Voice of the Customer for a Water Pistol.

Kids:
- Shoots Far
- Shoots Hard
- Shoots Straight
- Easy to Pull Trigger
- Trigger Doesn't Pinch Finger
- Fast Refill
- Easy Refill
- Looks Real
- Holds a Lot of Water
- Not Too Heavy
- Not Too Big

Adults:
- Lasts a Long Time
- Safe

Establishing Priorities

It is important to understand customer priorities relative to their requirements, because we generally cannot afford to devote substantial resources to all requirements. Some requirements will be more important to the customer than others. An importance rating, representing the strength of the customer's voice, should be assigned to each customer requirement, as shown for the water pistol requirements in the first column of symbols in Table 8–2. The organization can now begin to focus on important customer requirements. However, the focus becomes better tuned once a competitive analysis is added to the criteria for determining high-priority customer requirements.

TABLE 8–2
Product Planning QFD for a Water Pistol.

Symbol legend: ◉ Strong, ○ Medium, △ Weak.
Competitive analysis legend: □ Our New Gun, X Current Product, ○ Competitor 1, △ Competitor 2.

The measures are grouped under "Measures for Customer's Voice." The Competitive Analysis plot runs from Poor (left) to Good (right); overlapping symbols are shown combined.

Group	Customer's Voice	Strength of Voice	Shot Distance	Water Velocity	Shot Size	Shot Force	Aim Versus Result	Force to Shoot	Motion to Shoot	Time to Fill	Length	Width	Height	Weight	Strength	Reservoir Size	Looks Real?	Competitive Analysis (Poor → Good)
Kids	Shoots Far	◉	◉	◉	△	○		△										○ X/△ □
	Shoots Hard	◉	○	◉	○	◉		△										X/△ ○ □
	Shoots Straight	○	△			○	◉											□/X/△
	Easy to Shoot	○	△	△				◉	◉					△				○ X/△ □
	Doesn't Hurt Hand	△						○	△									□/X
	Fast Refill	○								◉						○		X ○ □/△
	Easy Refill	◉								◉								X/△ ○ □
	Looks Real	○									○	○	○				◉	○ □/X
	Holds a lot of Water	○			△					○	○	◉	◉	○		◉		△ □/X
	Not too Heavy	△									△	△	△	◉	△	○		□/X □/△
	Not too Big	△									◉	○	○	△				□/X
Adults	Lasts a Long time	○													◉			○ □/X
	Safe	◉															◉	□/X
	Target		30 ft	40 ft/sec	0.3 oz.	0.03 lbs.	Test Score 25	0.5 lbs.	1.4 in.	15 sec.	4.5 in	0.75 in	3.75 in	0.25 lbs.	X Modulus	6 oz.	Medium	

The results of a competitive analysis for the water pistol are shown at the right side of Table 8–2. The *O* and △ represent performance ratings of two competitors, while the *X* denotes the company's own performance on a current product. The square symbol indicates areas in which the toy manufacturer decided to focus resources. These areas represent requirements that are important to the customer and/or poorly achieved by the competitors. They give the manufacturer an opportunity to gain a competitive edge.

THE USER'S INTENT

Once clear statements of the customer requirements have been established, the user's intent in using the product can be determined (even before measures and targets are identified). It is very important that managers and engineers clearly understand what the customer wants the product to do. This information is important for developing a product that provides the appropriate function. In other words, the engineer should establish the intention of the product from the customer's point of view.

For example, suppose the engineer determines from the water pistol customer requirements that the function most critical to the customer is that the pistol shoot hard. A young girl wants to get her brother very wet when she squirts him with the pistol!

While the product concept selection and optimization activities will center around the primary intention, shoots hard, the other important performance requirements (such as shoots far and shoot straight) and important aesthetic requirements must also be considered. In fact, during the optimization activities, appropriate analysis methods are used for all of the important characteristics, as explained later in this paper.

In addition to the intention, engineers must understand the noise factors that the product will encounter during its use. Noise factors are factors beyond the engi-

engineer will want to select a product concept that has the greatest ability to resist noise. Such a product is said to be robust.

From watching and listening to water pistol users, the following noise factors were identified: distance from target, high or low target, and windy versus calm days. The engineer's objective is to select a product concept and to optimize the performance of the selected concept, such that the water pistol shoots hard on windy days and calm days, regardless of whether the target is nearby or far away, up high or down low.

HOW TO MEASURE CUSTOMER WANTS

In order for a company to know whether or not it is adequately meeting the important customer requirements, there must be some way to measure each requirement. Thus, the next step in overall product planning is to consider each important customer requirement and determine at least one corresponding way to measure it.

For example, the water pistol requirement "shoots hard" was evaluated by measuring the force with which the water hits a target placed a certain distance from the pistol. Measures for all the other water pistol requirements were determined in a similar manner, as shown in the top of the product-planning matrix (phase one of QFD, product planning) in Table 8–2.

The various customer requirements and their corresponding measures represent different types of product characteristics. Some relate to the product's primary function, while others deal with aesthetic characteristics, safety, and negative performance issues, such as performance in the presence of noise. The engineer must ensure that desired levels are met for all important product characteristics. However, the most important characteristics for ensuring function are measures of the product's primary function. These characteristics play an important role in selecting an appropriate product concept and in optimizing the product's performance.

In the water pistol example, shot force is the product characteristic used to measure the primary function related to the user's intent, shoots hard. Thus, shot force is the output used in defining the water pistol's ideal function.

Establishing Relationships between Wants and Measures

An identified measure (see the top of Table 8–2) may often be related to more than one customer requirement. Similarly, a particular customer requirement can sometimes be evaluated by several of the measures. To offer yet further guidance in establishing priority, the engineer considers the relationships between each customer requirement and each measure and denotes the strength of such relationships with symbols, as shown in the central portion of Table 8–2. The final product characteristics (see the top of Table 8–2) that strongly relate to a number of customer requirements are generally important characteristics for the engineer to consider.

Setting Targets

Once the important functional and aesthetic characteristics have been identified, the next step is to determine the target values that need to be achieved in order to meet or exceed the customer requirements. Such specific values are not required for a company's ongoing technology development activities or regarded in the initial optimization efforts, but they must be known in order to tune a particular product to meet the customer-specified requirements. For easy reference, these target values are generally placed in the appropriate columns near the bottom of the product-planning matrix (see Table 8–2).

PRODUCT CONCEPT SELECTION

It is frequently difficult for product design engineers to determine which design concept should be selected and developed. Concepts often evolve from considering customer requirements and the company's existing product technology. Using standard methods, such as Value Analysis/Value Engineering (VA/VE) and Pugh Concept Selection, a large number of concepts can be roughly evaluated and reduced to a few desirable ones.

Once the list has been reduced to two or three desirable concepts, the engineer needs an efficient and effective way to compare the candidates in order to select the one with the greatest potential. Parameter design experimentation using dynamic characteristics may be employed to evaluate the desirable concepts. This was done in the case of the water pistol. An ideal function was first established for each type of pistol.

Ideal Function and Signal-to-Noise (S/N) Ratio

The ideal function of a system refers to the desired relationship between an input signal (input that initiates

FIGURE 8–3
An Engineering System.

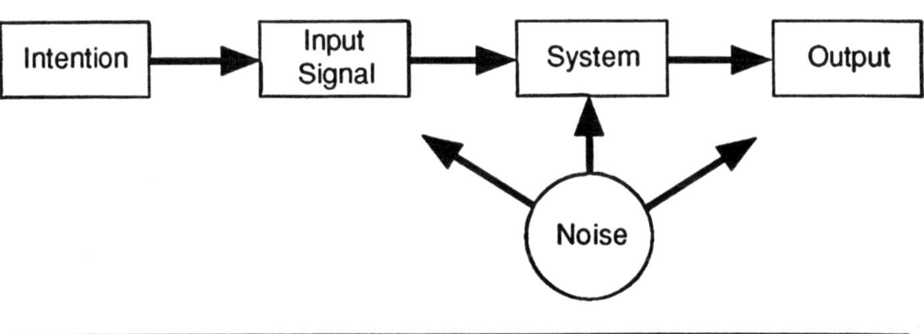

function) and the desired output (measures of the product's primary function). The system itself is composed of various components and design parameters that make the product work as intended. Figure 8–3 shows the essential elements of an engineered system.

First, the user's intention and noise factors in the field are determined by watching and listening to customers. Then, the primary functional output is determined from the measures of the customer wants. These three elements—intention, noise factors, and output—become the raw material for selecting an appropriate product concept. For each given product concept system, an appropriate input signal is identified.

Mathematically, the ideal function is represented by:

$$y = \beta M$$

where M = value of the input signal

y = value of the output

β = proportionality constant.

FIGURE 8–4
Ideal Function for a Trigger-Type Water Pistol.

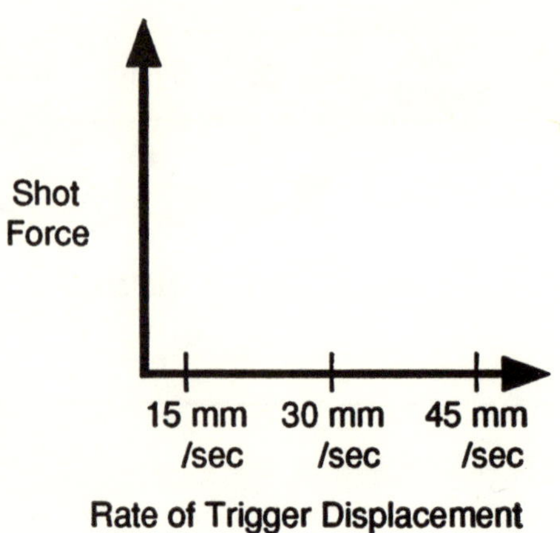

The intention of the water pistol is to shoot hard. The output, y, is shot force, and the major noise factors in the field are distance from target, high or low target, and windy versus calm days. An acceptable product concept is one that has the ability to consistently achieve the desired shot force in the presence of the changing noise factors over a reasonable range of input.

Two product concepts were considered for the water pistol: trigger-type (pull trigger to shoot) and handle-type (squeeze handle to shoot). For the trigger-type pistol, the input signal M needed to achieve the output (shot force) is the rate of trigger displacement (trigger displacement combined with how fast the trigger is displaced), as shown in Figure 8–4. Similarly, the input/output relationship is shown in Figure 8–5 for the handle-type pistol, with the input signal M being "rate of handle compression."

FIGURE 8–5
Ideal Function for a Squeeze-Handle Type Water Pistol.

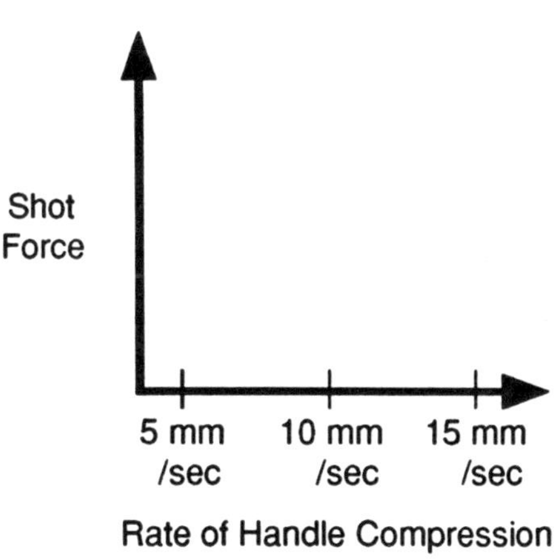

In a perfect world, shot force would increase proportionately as the rate of trigger displacement (or rate of handle compression) increases, and the input would be successfully transformed into usable output with no variation. But systems are engineered in an imperfect world. Nonlinearity and variation due to noise factors always degrade the output of our systems, as shown in Figure 8–6.

The useful portion of the output is the linear portion; the harmful portion is variation due to noise factors and nonlinearity. The ratio of useful output divided by harmful output is the Signal-to-Noise (S/N) Ratio. (When used to evaluate function, it is often called a dynamic S/N Ratio.)

The ideal function of an engineered system, described by the input/output relationship, serves as the basis for

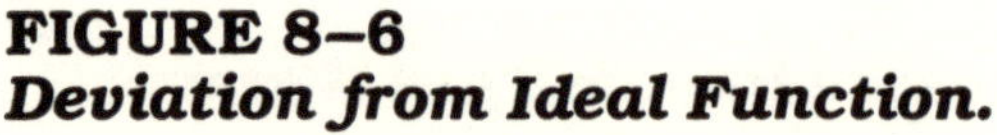

FIGURE 8–6
Deviation from Ideal Function.

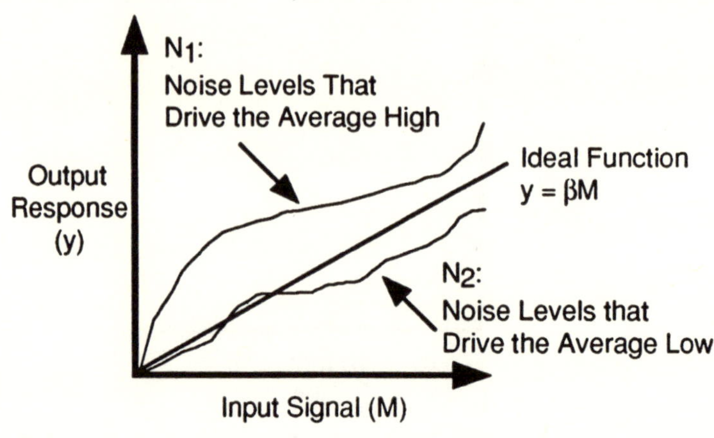

evaluating the technological level of the product. More specifically, performance is evaluated by the product's deviation from the ideal function. The S/N Ratio is a measure of this deviation.

Parameter design experiments (explained later) containing a few critical design factors were conducted on the water pistols, and the S/N Ratio was used to evaluate performance. Of the two pistols, it was found that the trigger-type pistol produced the higher S/N Ratio and therefore contains the greater potential to achieve the product's ideal function (see Figure 8–7). Thus, the trigger-type pistol was selected as the product concept to fully develop and optimize.

PRODUCT OPTIMIZATION

Once the product concept has been selected, the important product characteristics (measures of customer

FIGURE 8–7
Concept Selection.

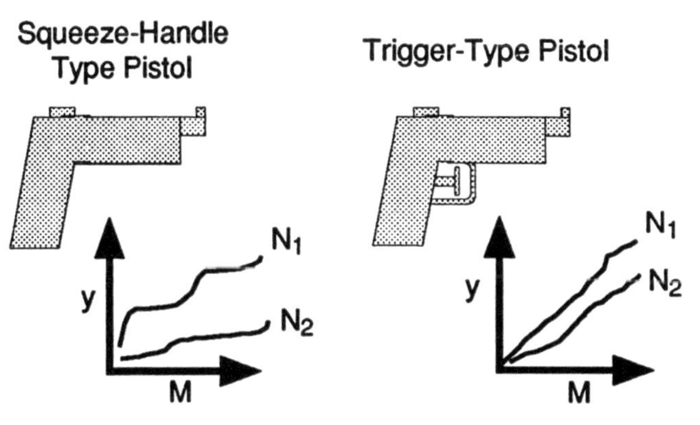

wants) and their targets are deployed to the second QFD phase, product development. At this stage, part characteristics (design factors) that can be used to attain the desired levels of the important product characteristic are identified. Some of these part characteristics will be included in the parameter design experiment conducted to optimize the product's performance.

Table 8–3 contains the product-development QFD matrix for the trigger-type water pistol, with the important measures (from the Phase I QFD matrix) and their targets listed on the left and the part characteristics across the top. The target values for the part characteristics recorded in the matrix were determined from a parameter design experiment and the allowable tolerances from a tolerance design experiment (explained later).

Parameter Design

Parameter design experimentation using dynamic characteristics is conducted to optimize the product's

TABLE 8–3
Product Development QFD on a Water Pistol.

Legend: ⊙ Strong   O Medium   △ Weak

Measures	Target	Importance	Check Pin: Specific Gravity	Check Pin: Ball Diameter	Housings: Inner Diameter	Piston: Diameter	Piston: Modulus	Spring: Rate	Lwr Check: Specific Gravity	Ball: Ball Diameter	Nozzle: Land Length	Nozzle: Inner Diameter	Nozzle: Tapered Lead In	Fill Cap: Fill Opening	Fill Cap: Snap Detail	Notes and Comments
Shot Distance	30 ft	O	△		O			△	△	△	△	⊙				Shot force consistent within distance
Water Velocity	40 ft/sec	O	Dropped: Same as Shot Distance													See Tech Report #TC-678
Shot Force	0.03 lbs	⊙	O		⊙		△	△		O		⊙				Input Signal to Water Pistol
Force to Shoot	0.05 lbs	O		△	O	⊙		O			O	O	△			Reduce Variability and Set on Target
Time to Fill	15 sec.	△												⊙	⊙	Develop New Fill Design
Target Nominal Value			1.1 g/cc	0.130"	0.75"	0.135"	170,000 psi	0.5 lbs./in.	2.2 g/cc	0.125"	0.100"	0.050"	0.020" x 45 deg	0.50" Diameter	Ref Dwg# 3298	
Allowable Tolerance			+/- 0.05 g/cc	+/- 0.002"	+/- 0.003"	+ 0.003 - 0.000"	+/- 5000 psi	+/- 0.05 lbs./in.	+/- 0.05 g/cc	+/- 0.004"	+/- 0.005"	+/- 0.001"	+/- 0.005" x +/- 5 deg	+/- 0.025"	Ref Dwg#3298	
Analysis Methods: Parameter Design		X	X	X	X	X	X	X	X	X	X	X	X			
Analysis Methods: Tolerance Design		X	X	X	X	X	X	X	X	X	X	X	X			
Analysis Methods: RFTA														X	X	

function. From such an experiment, an optimal combination of levels for design factors is determined such that maximal performance is achieved in the presence of noise over a specified range of input.

The design factors and levels evaluated in the parameter design experiment to optimize the trigger-type water pistol are given in Table 8–4. These factors,

TABLE 8–4
Product Design Factors and Levels.

Component	Control Factor	Level 1	Level 2
Check Pin	A. Density	1.1 g/cc	1.9 g/cc
	B. Ball Diameter	0.125"	0.130"
Housings	C. Inner Diameter	0.50"	0.75"
Piston	D. Diameter	ID+0.005"	ID+0.010"
	E. Flex Modulus	170,000 psi	220,000 psi
Spring	F. Rate	1.0 lbs./in.	0.5 lbs./in.
Lower Check	G. Density	1.3 g/cc	2.2 g/cc
	H. Ball Diameter	0.125"	0.130"
Nozzle	I. Land Length	0.050"	0.100"
	J. Inner Diameter	0.020"	0.050"
	K. Taper Lead In	0.020" x 45 deg	0.020" x 60 deg

A through K, were placed in an L_{12} orthogonal array (see Table 8–5).

For each of the 12 combinations of design factors, shot force measurements were obtained at each signal factor (rate of trigger depression) and noise level (N_1, N_2). A dynamic S/N Ratio was calculated for each test combination of the design factors. The average S/N Ratio for each level of each design factor was then computed (see Table 8–6).

The design factors having the greatest impact on shot force are factors A, C, D, E, G, H, and I, as shown in Table 8–6. (The larger the difference between the level averages,

TABLE 8–5
Product Parameter Design Experiment.

	A	B	C	D	E	F	G	H	I	J	K	M1		M2		M3		Sn
												N1	N2	N1	N2	N1	N2	
1	1	1	1	1	1	1	1	1	1	1	1	0.865	1.357	2.293	2.751	3.024	3.383	11.04
2	1	1	1	1	1	2	2	2	2	2	2	1.228	1.330	2.200	2.424	4.089	4.360	13.16
3	1	1	2	2	2	1	1	1	2	2	2	0.766	1.623	2.607	2.827	3.292	3.916	10.26
4	1	2	1	2	2	1	2	2	1	1	2	1.137	1.475	2.482	2.584	2.871	4.580	7.07
5	1	2	2	1	2	2	1	2	1	2	1	1.167	1.577	2.188	2.287	3.025	3.846	10.96
6	1	2	2	2	1	2	2	1	2	1	1	1.141	1.273	2.796	2.994	3.840	4.458	14.87
7	2	1	2	2	1	1	2	2	1	2	1	1.092	1.492	2.118	2.792	3.434	3.607	13.01
8	2	1	2	1	2	2	2	1	1	1	2	1.067	1.405	2.048	2.613	3.377	3.663	14.14
9	2	1	1	2	2	2	1	2	2	1	1	1.373	1.627	2.290	2.701	3.019	4.213	8.89
10	2	2	2	1	1	1	1	2	2	1	2	0.990	1.517	2.300	2.669	3.954	4.121	14.75
11	2	2	1	2	1	2	1	1	1	2	2	0.954	1.453	2.541	2.554	3.043	3.384	11.88
12	2	2	1	1	2	1	2	1	2	2	1	1.111	1.410	2.495	2.662	3.364	3.731	15.85

TABLE 8–6
Response Table: Dynamic Signal-to-Noise Ratio.

	A	B	C	D	E	F
1	11.23	11.75	11.32	13.32	13.12	12.00
2	13.09	12.56	13.00	11.00	11.19	12.32
D	1.86	0.81	1.68	2.32	1.93	0.32

	G	H	I	J	K
1	11.30	13.01	11.35	11.79	12.44
2	13.02	11.31	12.96	12.52	11.88
D	1.72	1.70	1.61	0.73	0.56

D = Difference

the greater the impact.) The best performance levels (with the higher S/N Ratio) for these factors are $A_2C_2D_1E_1G_2H_1I_2$. Hence, these levels were selected. Levels for the remaining factors were chosen on the basis of lower cost or other practical concerns, since their impact on the S/N Ratio was weak. When no practical concerns were present, the level with the higher S/N Ratio was selected. In this manner, an optimal combination of nominal values for the important design factors was determined to attain consistent, high performance in the presence of noise.

In addition to the dynamic characteristic shot force, desirable levels for other important product characteristics must also be determined. Even though Table 8–5 only shows shot force data and S/N ratios, data were also obtained (and corresponding S/N ratios computed) for shot distance and force to shoot. These two characteristics were treated as nominal-the-best characteristics in the data analysis. Best values were selected for product factors such that minimal variation is achieved in the presence of noise, and adjustment factors were identified to shift the mean on target. The final selection of product factor levels was based on the results of all three characteristics: shot force, shot distance, and force to shoot.

The last product characteristic deemed important, based on the customer requirements, is time to fill. The desired level for this characteristic was achieved by using Reverse Fault Tree Analysis.

Tolerance Design

Once the nominal values for design factors have been identified from parameter design, the allowable tolerances for these factors must be identified. To find out which factors' tolerances affect output variation and which do not, a tolerance design experiment may be conducted. The factors whose tolerances have a strong influence on the output variation may be tightened in order to further reduce output variation; tolerances with a mild impact on output variation may be relaxed. The allowable tolerances for the trigger pistol design factors, as determined from a tolerance design experiment, are given in Table 8–3.

THE MANUFACTURING PROCESS

An appropriate concept must be selected for the manufacturing processes by which the product is to be produced, as with the product itself. Information from a company's ongoing process technology development activities and knowledge of the product requirements are pooled to select high-potential manufacturing processes. These processes may be evaluated by means of the dynamic S/N Ratio in much the same manner as was described for comparing product concepts.

The key part characteristics and their targets (as determined from product parameter design) are deployed to the third phase of QFD, process planning. Process parameters and noise factors likely to have a strong influence on the key part characteristics are identified and included in a process parameter design experiment. Since a manufacturing process is dynamic (energy is being transformed), an ideal function is established for the process, and the dynamic S/N Ratio is used to evaluate function.

Once best nominals (targets) are determined for the process factors through parameter design, their allowable tolerances can be determined by conducting a tolerance design experiment, as was done for product design factors.

Finally, the key process factors and their targets are deployed to the fourth phase of QFD, production planning. At this point, production procedures are established in order to maintain the desired levels for the key process factors.

An important feature of the QFD process is the linkage from one phase to the next. QFD is a process; the input is the customer's voice, and the output is a set of practices and procedures for the production floor. The voice of the customer is translated into design parameters and deployed through product development, engineering, and manufacturing. This linkage helps ensure that the voice of the customer is captured in the final product.

SUMMARY

The product planning and development activities within a company must be driven by the customer's voice. Proper use of QFD (planning) and Taguchi Methods (optimizing) as means to develop products can ensure that the resulting product meets or exceeds customer expectations. Satisfying customers is a matter of survival.

REFERENCES

American Supplier Institute. *Quality Engineering, Dynamic Characteristics, and Measurement Engineering Five-Day Workshop Implementation Manual*, 1990.

______. *Quality Function Deployment Three-Day Workshop Implementation Manual*, 1989.

Taguchi, Genichi. *System of Experimental Dcsign*, vols. 1 and 2. Unipub, White Plains, NY: Kraus International Publications; Dearborn, MI: American Supplier Institute, 1987.

Chapter Nine

QFD Implementation at Chrysler: The First Seven Years

Robert J. Dika

ABSTRACT

Quality Function Deployment (QFD) appears to be a simple and rational method to translate customer requirements into appropriate company technical requirements. Although this is true, it is also the nature of QFD to challenge some of the basic assumptions and traditions of the new product-development systems in mature organizations. Since QFD demands a change in the ways that we think and act as a company, it will meet resistance.

This paper presents the story of successes and struggles that Chrysler Corp. has experienced in the integration of

Robert J. Dika is the quality planning executive—small-car platform, customer satisfaction, and vehicle quality office, Chrysler Technology Center, Chrysler Corp., Auburn Hills, Michigan. His paper is reprinted courtesy of Chrysler Corp.

QFD into its development process. Chrysler's story demonstrates that, although the road will be difficult, success is possible if those involved maintain their resolve and work within the current system. Chrysler's steady growth in both the number and significance of QFD projects over a seven-year period shows that QFD can be a strategy in the movement toward a Total Quality Management culture.

GETTING STARTED

Chrysler had net earnings in excess of 1.6 billion dollars and unit sales of over 2.1 million vehicles in calendar year 1985. Company officials referred to this as "reaping the harvest of our past efforts." Chrysler had emerged from the crisis and loan guarantees of the late 1970s as a much stronger company and exceeded even its own expectations for reestablishing itself. During 1985, Chrysler focused on diversification. It divided the corporation into four operational groups: motors, financial, technologies, and the newly purchased Gulfstream Aerospace. It launched a major joint venture with Mitsubishi called Diamond Star Motors. It committed to a five-year $12.5 billion product and capital spending program and added support to the advanced engineering organization, Project Liberty. On the product side, the H- (Chrysler LeBaron) and P-body (Dodge Shadow and Plymouth Sundance) vehicle lines were introduced. The company reaffirmed quality as the number one goal and stressed the necessity of being the low-cost producer.

The following year, Chrysler acquired American Motors Corp. (AMC). The integration of the two organizations required the attention of management at all levels. The company was able to achieve net earnings over $1.3 billion and sales remained over 2 million vehicles. The J- (Chrysler LeBaron Coupe) and C-body (Chrysler New Yorker and Dodge Dynasty) product lines were introduced.

During 1986, Chrysler made a significant push to improve the inherent reliability and consistency of its entire product line. Chrysler had survived from the brink of disaster and was starting to show that it could develop a record of product acceptance as well as profitability. But underlying this positive picture was a K-body-based car line-up with many fundamental problems and an aging stable of trucks. To improve its focus on quality, the corporation had licensed Phil Crosby's approach and was in the second year of implementing the Quality Improvement Process (QIP) companywide. At the same time, engineering was leading the effort to improve the technical capabilities of the product-development community. A number of new people with strong quality and reliability backgrounds had been brought in and were expanding the use of Failure-Mode and Effect Analysis (FMEA), design of experiments (DOE), Statistical Process Control (SPC), and other statistical methods within Chrysler.

During this time period, a group of us were working in an engineering office quality assurance function. Our job was to find and identify, through testing and inspection, product problems and push to get them fixed, often with only short-term effect. Our assignment was to keep the company in business while others were working to provide new products and manufacturing processes that could compete with the Japanese on the basis of quality.

On June 26, 1986, my boss and I were invited to the American Supplier Institute (ASI) offices to hear a four-hour introduction to a new technique discovered in Japan called Quality Function Deployment (QFD). We had no idea what it was about, but we were curious and there were a couple of openings ASI had recently become independent from Ford and was beginning to work with the other domestic auto companies, as well as the supply base, in applying some of the new ideas from Japan. This session was a freebie. Chrysler had participated in a seminar that was to include a discussion on QFD, but time ran out and ASI invited us back to complete the discussion. The Chrysler contingent included people from engineering quality and reliability, program management, Project Liberty, and the corporate quality office.

The people from ASI walked us through the fundamentals. They said that during their study missions to Japan they had encountered a variety of QFD matrices included in case-study presentations on other subjects and had inquired about them. They were just beginning to understand the potential of the technique. They talked about the voice of the customer and how through the House of Quality substitute quality characteristics were developed and deployed through each of four phases. They talked about matrix mechanics and how QFD was used in tandem with Taguchi Methods to identify critical design and process characteristics, optimize the product, make the design robust, narrow product variation about the target value, and, through proper training of operators, significantly improve quality.

But more importantly, they talked about transforming the quality culture. Larry Sullivan said that QFD unified the whole company in meeting the requirements of the new-product consumer and that "QFD represented a new way of thinking." He said that "QFD links quality technology together horizontally through the company" and that "it had been shown to reduce product development cycle time by one-third to one-half with a corresponding improvement in cost and quality." QFD was the technology of "prevention" in a manufacturing setting.

This seemed too good to be true. As quality assurance engineers, we were well aware that finding and fixing problems could not in the long run improve product quality. In fact, corrective actions may mask the inherent weaknesses in the design process and give the company a false sense of security. QFD seemed so logical. To my boss and me, it was a very compelling story.

When we returned and started to talk to the others, it was a little different. Although everyone from the Chrysler contingent liked what they heard, we were not sure how we should act on it. Program management seemed to say that it was good information, but they didn't choose to follow up on it. Engineering quality and reliability indicated that they had plenty on their plate and didn't know how they could get involved in promoting yet another technique. The corporate quality office was heavily committed and involved in developing QIP and not prepared to help. Since QFD is broad and systematic, the Quality Office may have actually considered QFD an alternative to QIP and thus thought its introduction ill-advised at the time. An engineer from Project Liberty and I agreed to attend an upcoming ASI three-day seminar to see what happened.

The ASI seminar on QFD began on July 7, 1986. Larry Sullivan and the ASI staff presented in detail what they knew about the subject based on their interaction with Japanese companies, but the principal speakers were Don Clausing from MIT and Akashi Fukuhara from the Central Japan Quality Control Association (CJQCA). The session began with a bang. Dr Clausing told us in no uncertain terms that "our present practice in the United States is a way to go out of business." He said that we could go out of business slowly or quickly, but the result would be the same if we did not change. He told us "the real reason that the Japanese have a cost advantage is that they have a better development process." His message was very clear. If we were not willing to study what they were doing and make breakthroughs in the ways that we develop products, we would not be able to compete.

On the afternoon of the first day, Mr. Fukuhara in a calm, assured tone walked us through the method. He had previously worked for Toyota Motor Co. as a quality assurance manager and later as a principal activist in their internal quality revolution of the 60s and 70s. He was now a consultant with CJQCA and was active in the training of the circle of Toyota companies, their supply base, and other CJQCA companies. Since Fukuhara was an engineer, his approach was both practical and real. Looking back on my notes from those sessions, I realize that he was cutting right to the heart of the matter. He was telling us exactly what we needed to understand.

He started by saying that "QFD cannot be done from a scattered point of view or with a scattered approach." It must be done systematically and must be integrated with the company's method of product development. He said that since we must find a method to improve, "QFD is a necessity for the future."

He said, "Solve problems at the design and blueprint stage. Prototypes are to prove that the product works. . . New model time is the time to solve problems. . . Identify the key items such that all areas of the company can focus their efforts and create significant impact on both quality and cost reduction. . . These

items only require QFD." He said that "QFD is an itera-
tive process, a series of finer and finer mesh screens.
First, look for big problems, then finer and finer."

He also gave us a way to use QFD as a countermeasure
for current problems. He said.

> First, look at conditions of the market place. Then look
> where you are . . . Tear-down of competitive vehicles is es-
> sential . . . Determine where you are insufficient or where
> competitors are much better . . . This results in an absolute
> need to find a solution . . . All areas of the company must
> cooperate . . . QFD is to clarify objectives, unify all areas
> on the same path, and assure smooth quality transmission
> (communication).

He said that quality assurance should have three major functions:

1. Use QFD to improve methods and involve all departments. "Quality management's role is to identify the key items" and "QFD is a system to involve
   all people in the process of improving them . . . QFD
   is the basis for a network of cooperation to be
   formed . . . Use QFD to train young engineers to
   think beyond themselves."
2. Reliability is the responsibility of the design manager. "QFD is the responsibility of managing engineers; if it is the Quality Assurance department's
   responsibility, it will not work."
3. Existing quality issues must be solved when the
   next model is introduced. "Inspection is not to improve quality. It is to determine what is good and
   what is bad . . . Quality does not cost money. It
   saves money . . . Saving money motivates people
   at Toyota."

He said to get started. "Start with a small problem,
learn from it, work your way up—don't start with life

and death issues.'' Show how ''conflicting design characteristics can be optimized analytically using parameter design.'' Then ''present good case studies in public forum,'' and the success stories will cause QFD to grow.

Well, this hit me like a ton of bricks. The Japanese weren't successful because they were culturally different, had better workers, or even had economic advantages. They were successful because they were systematic and behaved sensibly as an organization. They operated as a team focused on necessary results and the keys to success.

I returned to work fully expecting that everyone else would be as excited about this discovery as I was. No such luck. Nobody seemed ready to take on the task of selling the technique to the company. We decided to just go ahead and start ourselves. My boss and I put together a short pitch on QFD and started making presentations wherever we could get an audience. We selected people who were either influential and had reputations for their willingness to try new things or, in our estimation, were prime candidates for application.

From the beginning, our strategy had two elements:

1. Awareness. Let people know that QFD exists and how it can be applied. Look for people interested in starting projects.

2. Development of case studies. Begin application of QFD to build knowledge and create success stories that will attract others to use it.

Since we didn't have many takers from the engineering departments or other line organizations, our first project was one where our organization had lead responsibility. The topic was ''Underhood Corrosion and Splash Protection.'' Our first QFD matrix appeared in August.

During the second half of 1986, we continued what we had started. We continued making house calls on managers and executives around the company, telling our story and looking for interest. On September 30, we invited ASI to help us officially launch our effort. More than 50 influential managers and executives attended a four-hour seminar on QFD principally conducted by Larry Sullivan and John McHugh of ASI. Don Clausing also participated. In addition, we continued to look for pilot projects.

By December, we had raised enough interest that we were invited to make a presentation to what was called the Advanced Product Committee (APC). We viewed this as a big opportunity to get management support. The APC included the executive vice president of product development; the vice presidents of the engineering office, product design, and program management; the Dodge, Plymouth, and Chrysler brand general managers; and the director of corporate timing.

The session went well. During the meeting almost everyone voiced support for proceeding. But there was also a voice of caution. During the previous year, there had been a significant effort to increase the use of FMEA, which had resulted in a proliferation of FMEA projects. Many were initiated primarily to please management and pad the number of projects started. The APC decided to endorse the use of QFD and gave us the go-ahead to proceed, but it also decided not to support the effort too visibly in order to avoid false starts or overreaction on the part of the organization. This was not exactly what I wanted to hear, but we came out of that meeting with a clear OK to proceed. I believe that they saw our enthusiasm and wanted to let us give it a try.

By year end, we had stimulated four QFD teams within engineering and Project Liberty had initiated one.

YEAR ONE

In 1987, Chrysler earned more than $1.2 billion, and with the addition of the AMC product line, sold more than 2.2 million units. The integration with AMC was beginning to take form and the Eagle brand was created. The corporation sold its interest in Peugeot that had been acquired with the AMC merger. The president of the corporation retired. The BB-body (Eagle Premier) product line was introduced. Corporate spokespeople indicated that "the economy is starting to squeeze."

With the endorsement of the APC and the support of local management, we increased our efforts in promoting QFD. Our goal was to develop our expertise and to build on our experience. We were still conducting informal discussions with interested people in their offices, but we were also broadening our approach. In the first quarter, I began teaching a two-hour QFD awareness class through the existing engineering quality and reliability training program. Our intention was to introduce people to the concept on a wider scale and to identify individuals interested enough to form a team. We found that almost everyone had a positive reaction to the concept, but relatively few felt that they could step forward as a team leader. Since there was no formal directive to use QFD, any projects started required someone from the line organization to become a "champion." We were plowing new ground and everyone felt that they were already very busy.

Although these awareness sessions have changed over the years, we continue to offer them on an open-enrollment basis. There will always be new people that discover the term QFD and want to find out what it is.

We had to develop nearly all of the materials that we used ourselves. Although ASI was preparing a manual and other companies were doing the same kind of thing that we were, there were no text books or definitive materials that met our needs. The first substantive book on the subject, *Better Designs in Half the Time,* by Bob King of GOAL/QPC, wasn't available until late in 1987. *The Customer-Driven Company—Managerial Perspectives on QFD,* by Bill Eureka of ASI and Nancy Ryan, wasn't published until 1988. Examples like the Toyota rust study, although good for specialists to study, were not attractive to the engineers and managers that needed to be leading if QFD was to be effective. It took us many years to develop simple, clear, yet motivating materials that had sufficient detail to support teams getting started.

At the same time that we were developing our own materials, we were supporting ASI in developing theirs. For a period of time, I was making an appearance at each ASI three-day facilitator seminar and giving a testimonial on its potential and our support at Chrysler for suppliers that wanted to pioneer its use. Through our interaction with ASI, we were able to form alliances with several interested supplier companies. Our staff and the supplier quality and reliability staffs were able to join forces and initiate collaborative projects. Together, we were able to stimulate, facilitate, and train a number of new QFD teams.

Although we each had our own concept of QFD, none of us still had much experience. We had to be active within the project teams to gain experience and discover how it really worked. We were all learning, teacher and students alike. Looking back on it, we were well on our way but still had a lot of incomplete notions about the details of the technique. Everyone involved in QFD at the time needed people to talk with who were experiencing the same problems.

We persuaded our local management that I needed to see QFD in action and to talk to the real experts. I was able to become a member of the ASI study mission to Japan in April of 1987 and again in 1988. These trips to Japan were wonderful learning experiences. Our hosts were gracious, supportive, as well as candid. They were willing to share anything that we wanted to know short of specific, proprietary product details. The time spent with people from Japanese companies who really knew what they were doing and with executives from other US companies who were struggling with similar issues was invaluable in developing our understanding. Having visited Japan, I was also able to discuss the good examples that I had seen and thus acquired some credibility as an expert within my company.

That spring, we became involved in a project that became a major boost to the overall QFD implementation effort. The product-development community was working very hard to identify current product shortcomings and were making plans to improve the customer acceptance of all product lines to best-in-class levels. On the subject of the P-body (Dodge Shadow and Plymouth Sundance) there was a lot of activity. Meeting after meeting was held at every level, from working engineer to top management. Each functional organization had a plan to make improvements that they thought necessary. There was a marketing plan, an engineering plan, a manufacturing plan, and so forth. Since it was clear that there were limited resources available to execute all of these sometimes conflicting plans, it was suggested that QFD be used to sort out the overall corporate priorities. A group of reasonably high-level executives was assembled to "do a QFD."

The resulting world-class P-body QFD set a standard and developed a model for other groups to follow. Due to the credibility of the people involved, the concept of QFD gained credibility. The world-class P-body team completed what is now known as a preplanning matrix. The customer requirements were listed with the aid of some existing qualitative market research and then prioritized using quantitative things-gone-wrong, marketing, and syndicated competitive benchmark survey data. The team concluded that five new QFD teams needed to be formed to deploy the critical customer needs. The use of the QFD preplanning matrix for the P-body became a model for upcoming new model programs to follow.

During this first year, we found ourselves meeting and working much more frequently with people from marketing and corporate research. When an engineer starts to think about what the customer wants and needs, he or she naturally starts to look for data. We found that most often there wasn't definitive or detailed voice-of-the-customer data available, so we began to utilize existing market research clinics, which were traditionally used to evaluate styling, image, or positioning only, to ask additional "QFD questions." We found the corporate research organization supportive, but developing opportunities and methodologies for acquiring valid qualitative and quantitative data to support QFD teams demanded much time and effort.

After just a year of trying to introduce QFD at Chrysler, we were asked to present a paper at a joint American Society for Quality Control (ASQC) and ASI "Conference and Workshop on Taguchi Methods and Quality Function

Deployment." Needless to say, this was a bit premature, but quality professionals were anxious to see any actual experience in implementing QFD in the United States. We reported the following: QFD gives us the opportunity to talk to our colleagues about the process of product development rather than only talking about product problems. This takes us from the realm of problem solving to problem prevention. We said that our strategy was simple. First, we wanted people to become aware of QFD. Second, we were starting to learn by doing. QFD seemed to have both benefits and barriers. On the positive side, we reported that it helped us focus on the customer, improve communication, prioritize work, team build, understand other organizations, be more systematic, involve management, identify conflicting requirements, and identify component interactions. On the negative side, we reported that it involved a time commitment and that there was a learning curve, that some felt it was too complex or not clear, that it was difficult to get customer input, and that the hard benefits were unsure. We also reported that progress was slow, that sometimes teams lacked patience and full participation, and that it may not result in quick improvements in the product.

We said that we were going to keep working to enumerate the benefits and knock down the barriers. We felt that QFD had staying value and that it did what it was intended to do. It was up to us to make it happen. We asked the attendees to become students of QFD because it was an investment in their futures.

Later that year, we were able to move two additional engineers part-time from problem-solving activities into facilitation of QFD projects. During the year, six new QFD projects were started and we were facilitating nine active projects at year end.

YEAR TWO

In 1988, the corporation recorded a solid year with earnings of $1.05 billion and more than 2.5 million vehicle sales. The corporation reported the "smooth integration of AMC operations" and announced the commitment to build a new billion-dollar technology center in Auburn Hills, Michigan. The A-body (Dodge Spirit and Plymouth Acclaim) and Plymouth Laser sport coupe from Diamond Star were introduced.

We had shown some success in applying QFD and achieving the support of those people directly involved in QFD projects. On a couple of occasions, we were able to review the projects with our executive vice president, and he was beginning to be more active in supporting the formation of new teams. But results were still soft; improvements in customer satisfaction with the product systems that we were studying would not be apparent to the customer for some time. In the QFD group, our goal was to develop stronger, more influential projects and to show hard results.

Early in the year, we began working on a total-vehicle project, code-named LH, and planned for its introduction as a 1993 model. A couple of individuals in the LH program-management organization felt strongly enough about the merits of QFD to get started. The team included people from program management, the product design office, vehicle engineering, process engineering, finance, engineering quality assurance, and one executive from each of the brand management organizations (Chrysler, Dodge, and Eagle).

Since this was to be a significant product for Chrysler, funds had been set aside to conduct early market research to identify wants and needs. Team members actively participated in these research events, which were video-taped for use by the product community later. The team utilized transcripts from these events and the affinity diagram technique to organize the information. Then following the pattern used by the world-class P-body team, they utilized the QFD preplanning matrix to display quantitative data and establish overall priorities for the vehicle line. This transpired over an eight-month period prior to program approval.

The team recommended that five new QFD teams be created to study the critical product systems and establish technical functional objectives: powertrain excellence, smooth chassis, image, quietness, and interior package. An executive was identified to sponsor each team, and at least one member from the total vehicle QFD project was assigned to each of the five teams to provide continuity and leadership. Each of these projects was a landmark for us. They involved principal people from the line departments at the working level, and each project proceeded well into the design phase of QFD. Some were documented more rigorously than others.

About this time, the engineering office was reorganized and our quality assurance organization became part of program management. The two engineers who had been working with me as QFD facilitators were reassigned. Three engineers, two with backgrounds in reliability and one with a manufacturing background, and I formed what we called the QFD systems group within program management to strengthen the QFD effort. This discontinuity in people forced us to go back to the beginning, rethink our mission and objectives, and train the new people thoroughly in QFD. The new people attended both the ASI and GOAL/QPC facilitator training seminars and immersed themselves in the examples and materials available. They also followed the lead of the more experienced facilitators until they were prepared to go out on their own. The training and development of these new people consumed quite a bit of time, but this investment more than paid off over the next couple of years.

During the year, the corporate quality office started supporting our facilitation effort by funding outside training. We were able to supplement our facilitators by bringing in ASI instructors to do customized half-day or full-day sessions of classroom training. ASI was better prepared than we were to develop high-quality training, and it relieved us of the responsibility to do it. This provided another big boost to our effort. The vice president of the quality office also started more actively speaking out in support of QFD.

This growth in QFD brought with it its share of problems. I am not exactly sure how it started, but there were occasional reports that one or more corporate officers were against QFD. Certainly there were feelings that there had been time and money wasted in the voice-of-the-customer market research that had become associated with QFD. There was also a feeling that QFD took a long time and produced only common-sense results. On October 18, we were asked to meet with the president and vice presidents to discuss how QFD should be used "intelligently."

We reported that QFD by its nature identifies what is critical and is self-prioritizing. We said that QFD had gone a long way in building buy-in of customer requirements and improving teamwork. We noted that we were beginning to work together, that QFD was growing slowly, and that we would continue to use QFD intelligently. Although they were cautious in their support, once again top management agreed that we should continue.

On November 15, 1988, a large-scale meeting was conducted to officially kick off the five new LH QFD teams. Our executive vice president, who was beginning to be more active in sponsoring our activities, retired late in the year, so the new vice president of vehicle engineering was asked to speak to the group in support. During his remarks, he gave us two more pieces of advice. First, he said to focus on the LH. That was our goal: to make the LH right for the customer. He said that he didn't really care if we created a lot of Houses of Quality or if we filled in all of the boxes in all of the rooms, although he was sure that if we did we would end up doing a very good job. But he wanted us not to get too hung up on the methodology. Second, he said that we should not look for him to tell us how to do our jobs. It was the LH team that was responsible and they should decide what were the right methods to use. He said that he wanted us to keep making progress and that he had confidence that we would do well.

During the year, 14 new QFD projects were started, and we were facilitating 17 active projects at year end.

YEAR THREE

During the next year, the five LH spin-off teams became involved in the second phase of market research and were principal participants in a research project held in Chicago. Many team members were able to go to Chicago, ride in vehicles with customers, and talk to them. They were able to ask questions about the customer's perceptions. Some of the vehicles were instrumented to sense vehicle performance on a real-time basis as the customers were driving and responses recorded. The technical measurements were subsequently correlated to the responses made during portions of the test drive. This event was a positive experience for all of the teams and in many ways set the tone for LH development. These teams carried on with QFD into the design phase and were able to show in subsequent research events, prior to production, that their efforts were successful in providing performance that excited their customers.

Early in the year, there was another reorganization and the QFD systems groups were made part of the vehicle engineering staff organization. We were separated from our long-time boss, who had supported us from the beginning. We found ourselves reporting to a new director. Luckily, he was very supportive. He had traveled with me on the study mission to Japan in 1988 and had seen QFD's potential. The QFD systems group was renamed quality planning as a means to keep the QFD effort protected.

In May of 1989, Chrysler started to make cutbacks. Although able to report $359 million in net earnings and more than 2.3 million vehicle sales worldwide for the year, the corporation announced that we needed to start "to take tough, painful actions to get trim for the downturn." The chairman announced the goal of cutting $1 billion from the corporate cost structure by 1990. The vice chairman left the company.

Even in the face of cutbacks, we were able to bring Fukuhara in to talk to our teams three times in 1989 and once the following year. Not only were we able to have a wide range of team members hear him speak and advise them on their projects, but we were able to have him consult with some of the forming platform management teams. He helped us focus ourselves on the critical elements of QFD. He was a wonderful teacher for us and a terrific boost to our efforts.

Although we were still developing new and better QFD projects, things were getting rougher. Rumors persisted that QFD was not fully supported by top management. It was difficult to find people to be QFD pioneers. To dispel some of the rumors, the president distributed a letter in June that said, in part: "QFD is a resource-intensive effort and has to be used selectively where it will produce the most results. I strongly support its continued application where you feel it's appropriate." In July, during a training session on QFD for corporate officers, he repeated that there seemed to be a misunderstanding. He said, "QFD is a powerful tool . . . Use it where it can truly make a difference . . . It has got to be used intelligently . . . Misapplication will destroy it, if we are not careful."

Although top management supported QFD, they were not active in its support and, for a while, getting new QFD projects started was very difficult. We did have strength in the middle management ranks, however. During this period, one executive stated, off the record, "We are still going to do QFD, but we're not going to tell anybody."

In December, we began working on a total-vehicle project for a new small car, code-named PL. It would eventually become the Dodge and Plymouth Neon. My original boss had been assigned as program manager of the PL and wanted our help. With a sponsor in an influential position within a platform organization, this project would prove to be another big opportunity.

The PL project was something very different. This was Chrysler's "little car that could." Few really wanted this car to happen. The chairman certainly had concerns that any US company could design and build a small car in North America at a profit. But the pressures of government Corporate Average Fuel Economy (CAFE) standards made a small-car strategy a necessity in order to sell larger, more profitable car lines.

Corporate planners were evaluating dozens of alternatives, mostly based on a P-body-minus approach. This is where limiting product features and drastically lowering the price would be calculated to significantly increase sales volume. None of these alternatives seemed to make real sense. Finally, the platform was given the chance to propose an alternative. The small-car team saw their challenge: Develop a car (1) at L-body variable cost and weight (Dodge Omni/Plymouth Horizon), (2) with P-body size and features (Dodge Shadow/Plymouth Sundance—a bigger vehicle), (3) that the customer saw as best-in-class in those attributes that were really important, and (4) that could be delivered as a 1994 ½ model within planned investment constraints.

The small-car project had something else that was different. Within program management, there was a small group of relatively young, forward-thinking people at the working level who were of a like mind. They truly believed that the new small car had to be customer-driven. They were process-oriented. They knew that this project would have to evolve over a period of time and many aspects would have to be managed. They knew that it wasn't going to be easy and that they were going to have to be teachers of a new way of doing business. Most importantly, they believed in self-managed, empowered teams as the only way to be successful. If this wasn't a belief in QFD, I don't know what QFD is.

We became members of this core group of "original PL'ers." On the PL project, it wasn't just the three of us trying to promote QFD; it became a group of six or eight. Early in the program, this team started meeting often. Since the company was trying to save money, we held what we called "on-site/off-site meetings" to discuss strategy. We would order pizza, bring in bottles of pop, bags of chips, and macadamia nuts and seal ourselves off in a conference room for an afternoon to strategize. We would work out what needed to be done during the next interval.

Over time, this core group was able to gain credibility and influence the management team of the small-car platform, but more importantly, they were able to work directly with the many small teams that were created to execute the new PL.

During June our quality planning group participated in the first Symposium on Quality Function Deployment, which introduced us to many new people from across the country in a variety of industries that were applying QFD. We continued to be active in this ongoing annual symposium from 1989 to the present and found it to be a source of information and support. During the year, we were also able to acquire some personal computer-based software for producing laser-printed QFD charts. This was attractive to some of our engineers and helped us to reduce the hassle of drawing the matrices by hand or with computer-aided graphics packages.

During the year, 16 new QFD projects were started companywide, and we were facilitating 17 active projects at year end.

YEAR FOUR

In 1990, Chrysler's 65th anniversary, the corporation reported the "fiercest price competition on record, a recession, and the menace of war." The industry was absorbing huge incentive costs and North American car and truck capability exceeded demand by 5.2 million units. Corporate earnings dropped to $68 million and sales to less than 2 million units. The Supplier Cost Reduction (SCORE) program was initiated to work with the supply base to find major cost reductions, and the chairman raised the bar on the corporate cost reduction effort to $3 billion by 1991.

The corporation affirmed that it was a car and truck company whose mission was to satisfy its customers. Gulfstream and much of the company's holding of Mitsubishi stock were sold. The corporate strategy was to have the highest possible quality *and* the lowest possible costs. Chrysler was determined to continue to adapt to the changing demands of our customers and the dynamics of the global automotive market.

In January, vehicle-development activities were officially reorganized into four "platform" teams: large car, small car, minivan, and jeep/truck. The quality planning group was reassigned to engineering scientific affairs, and we began reporting to yet another director. Our new director, again, was supportive and the PL project became our prime focus.

We knew that we had not been using QFD as efficiently as we should, so we established a goal to shorten the time that it took to get results from a QFD project. We also wanted to push into the downstream phases, and, since we were now a quality-planning group, to reach beyond our earlier limits and try to integrate the steps of QFD into overall platform work plans. To improve our facilitation skills and to make us more interchangeable, our quality-planning group worked together to develop an internal QFD manual. The QFD manual was used to add detail and Chrysler terminology to the concepts presented generically in the ASI training materials.

Like the LH, we were able to establish a meaningful preplanning matrix for the PL, which spun-off into five second-level QFD projects: reliability, responsive powertrain, structures, interior/ergonomics, and fun-to-drive. In all, 20 PL QFD projects were initiated. Many of these drew heavily on the work of earlier LH teams. In addition, we began to work directly with the PL management team to improve their leadership of the overall program. The PL was the first program where QFD was used as a primary strategy. Although LH was influenced by QFD, it happened through the individual efforts of team members and leaders. On the PL, QFD became part of the system.

Mid-year, quality planning was reassigned to engineering operations, and we began reporting to the executive engineer of engineering quality and reliability.

Late in the year, a preplanning matrix was developed for the NS/GS minivan program, the renewal of the Dodge Caravan and Plymouth Voyager planned for the mid-90s. Since this was an important product for Chrysler, it was supported like the LH with a substantial amount of up-front customer research. This preplanning matrix once again established the overall product priorities of a total vehicle program and influenced program direction in a meaningful way. Several system-level QFD projects were spin-offs. We were beginning to develop a pattern where all new model programs utilized QFD preplanning to establish broad priorities and direct extra effort into the critical areas.

This was a rough year for us, but we continued to be able to facilitate QFD teams at the working level. Although by the end of the year cutbacks were being felt and our outside training resource was eliminated, 21 new QFD projects were started and we were facilitating 22 active projects at year end.

YEAR FIVE

In 1991, Chrysler lost $795 million on less than 1.9 million unit sales. We were sticking to the $3 billion cost-reduction program and sold the 50 percent holding in Diamond Star Motors. On the product side, the improved AS-body (Dodge Caravan and Plymouth Voyager) was introduced, and the new Jefferson North assembly plant in Detroit was opened. The first groups began moving into the new Chrysler Technology Center (CTC). A new quality initiative called business process redesign was established at the corporate level, and we began to get involved in this effort. At the same time, we still wanted to emphasize the production phases of QFD.

The reorganization into platforms had created conditions that allowed the small-car platform the freedom to incorporate many new ideas. Certainly, this platform made a real commitment to utilize crossfunctional teams extensively at every level. During the planning and design phase, a network of "product teams" was created to manage all of the product systems. These teams were substantive, officially chartered forums for all product goals to be discussed and managed. Often they would charter smaller "work teams" to execute product-development tasks and activities. QFD doesn't work well without a team. The creation of these working-level teams provided a fertile opportunity to grow QFD projects.

A crossfunctional team at the executive level called Synthesis was created to provide direction to the product teams. Their role was to assure that all program objectives, including quality, cost, weight, investment, and timing, were rationalized at the vehicle level. They believed that all objectives needed to be in harmony at any point in time. All objectives needed to be achieved simultaneously if the program was to be successful; therefore, they had to be managed at the same time. Synthesis members continued to meet at length, every week, to hear reports from the product teams on a rotating basis and to resolve issues related to the product-development process. These meetings were the opportunity for the executives to lead, manage, and train their people. It also provided a means to measure progress at major program milestones.

Synthesis also sponsored training for the product teams in many subjects, including QFD. This training, which was developed from within the platform by platform team members, emphasized a practical approach. It was delivered on a just-in-time basis as the teams started the activities. This was a departure from conventional, open-enrollment training that had been the norm in the corporation. The training reflected an integration of the technical aspects of tool application with management direction and policy. Although the teams were free to use any methods they wanted, guidelines were presented and supported. This support and involvement by Synthesis grew stronger as the program moved through time.

Due to our involvement on the PL core team, we had plenty of substantive work to do. Our emphasis shifted from a focus on QFD projects to an involvement in developing an overall quality plan within the platform structure and the integration of QFD activities into overall product-assurance plans. We were helping people understand how to utilize the whole toolbox of quality and reliability methods within the timing of a new-product program. We focused on the goals for product function, reliability, and process capability. We also started to better understand how quality-related activities had to be integrated with activities primarily driven by other considerations. (On one of his visits, Fukuhara had told us that at Toyota 70 percent of an engineer's time was spent on creative cost reduction.) We were working to understand how we had to apply QFD if it was to achieve the results that were associated with it in Japan. Quality activities could not be separated from other program activities. We were finally understanding how cost, cycle-time reduction, and quality improvement had to work together.

Working with people from program management, we started to study the books of Henry Ford, Shigeo Shingo, Kaoru Ishikawa, Yasuhiro Monden, and Taiichi Ohno. We began to integrate our understanding of the Toyota production system and Just-in-Time (JIT) manufacturing into our overall quality plans. A program to expand the use of Design for Manufacture and Assembly (DFMA) was also ongoing.

During this period, we also started to discover the usefulness of Stuart Pugh's Concept Selection Matrix. The House of Quality was considered a diversion by many engineers. It took them awhile to discover why they needed to do it. But we found that engineers were immediately attracted to the Pugh Concept. It provided a means to solve problems for which they were directly responsible: selecting the most appropriate design or process concept. The Pugh Concept is a simple tool and its very simplicity is what makes it so powerful. It is not restricted to external customer requirements but allows the requirements of every internal function to be recognized. Teams were immediately able to see results and the use of the Pugh Concept grew very quickly once we started to actively promote it.

Mid-year, the engineering quality and reliability organization was transferred to the corporate quality office. Two of us who were doing quality-planning work found ourselves in a newly created platform quality-planning group, and one of us remained in vehicle engineering. The charter of our new group was much like what we had been doing already, but now there were more people at a higher grade involved. Since the new organization was an official quality-planning group, we were pulled even farther from being exclusively QFD people.

Late in the year, preplanning projects were completed for the mid-90s renewal of the ZJ (Grand Cherokee) and XJ/YJ (Wrangler and Cherokee) models within the jeep and truck platform. We saw the pattern that had been successful for us in the car and minivan platforms begin to happen in the truck area.

During the year, 19 new QFD projects were started and we were still facilitating 28 active projects at year end.

YEAR SIX

In 1992, the economic cycle started to reverse. The corporation was able to earn $723 million on more than 2.1 million units. Product programs that were maintained during the downturn and the effects of corporate cost-reduction efforts started to pay off. The new Dodge Viper, ZJ (Jeep Grand Cherokee), and LH (Chrysler Concorde, Dodge Intrepid, Eagle Vision) models were all introduced. Our group and the major portion of the corporation's product-development organizations were finally co-located at CTC.

In January, the PL project was moving from an emphasis on design and engineering to manufacturing. Thirty new "process teams" were created with the goal of developing shop-floor control plans for the critical systems. The process teams were a natural extension of the earlier product and work teams, but the emphasis was different. More people from manufacturing and the assembly plant became active and started to take leadership postures. As part of the PL core team, we were called upon to both facilitate and develop training for these new teams. The JIT training format that we had developed earlier was utilized to provide guidance for the new process-oriented work. This allowed us to more actively promote the third and fourth phases of QFD.

Working with talented people within the small-car platform, we were beginning to apply not only QFD, but the concepts of source inspection and mistake-proofing based on the work of Shigeo Shingo that we had been studying. We also began to more fully utilize Process Failure Mode and Effect Analysis (PFMEA) and control planning (checklists), which were always part of the QFD arsenal. We were trying to eliminate the need for Statistical Process Control (SPC). This work helped us mature our understanding of how to utilize QFD in a manufacturing and assembly environment and finally push through to the final phase.

Mid-year, there was another reorganization. Our small, unified core of QFD facilitators was finally separated completely and assigned to varied duties as part of the platform quality-planning function. What we had learned in studying and applying QFD over a six-year period was now being integrated into the overall process of quality planning. A major initiative was started in jeep and truck platform to adapt the ideas that were becoming recognized as successful in the other platforms to the much different environment in the truck organization. Within this initiative, the elements of QFD became part of the guidelines for their newly formed product-assurance teams (PAT).

During the year, 14 QFD projects were initiated, with the total coming to 95 projects that we were able to facilitate over the six-year period.

DISCOVERIES

There has been steady progress at Chrysler in the use of QFD (see Figure 9–1). Although half of our projects dealt with only the first phase, we have explored many of the other aspects of QFD. We have utilized matrices in each of the four phases described by ASI. Of the 30 types of matrices represented by the GOAL/QPC Matrix of Matrices, we have explored 12 of them to some degree or another. We have also created matrices of our own (see Figure 9–2).

Every year the number of active QFD projects has increased, but more importantly, the significance and influence of succeeding projects have also grown. The body of this paper has attempted to show how this has happened.

In retrospect, we know that we did some things right and some things wrong. We probably expected it to be

FIGURE 9-1
Summary of QFD Progress at Chrysler Corporation.

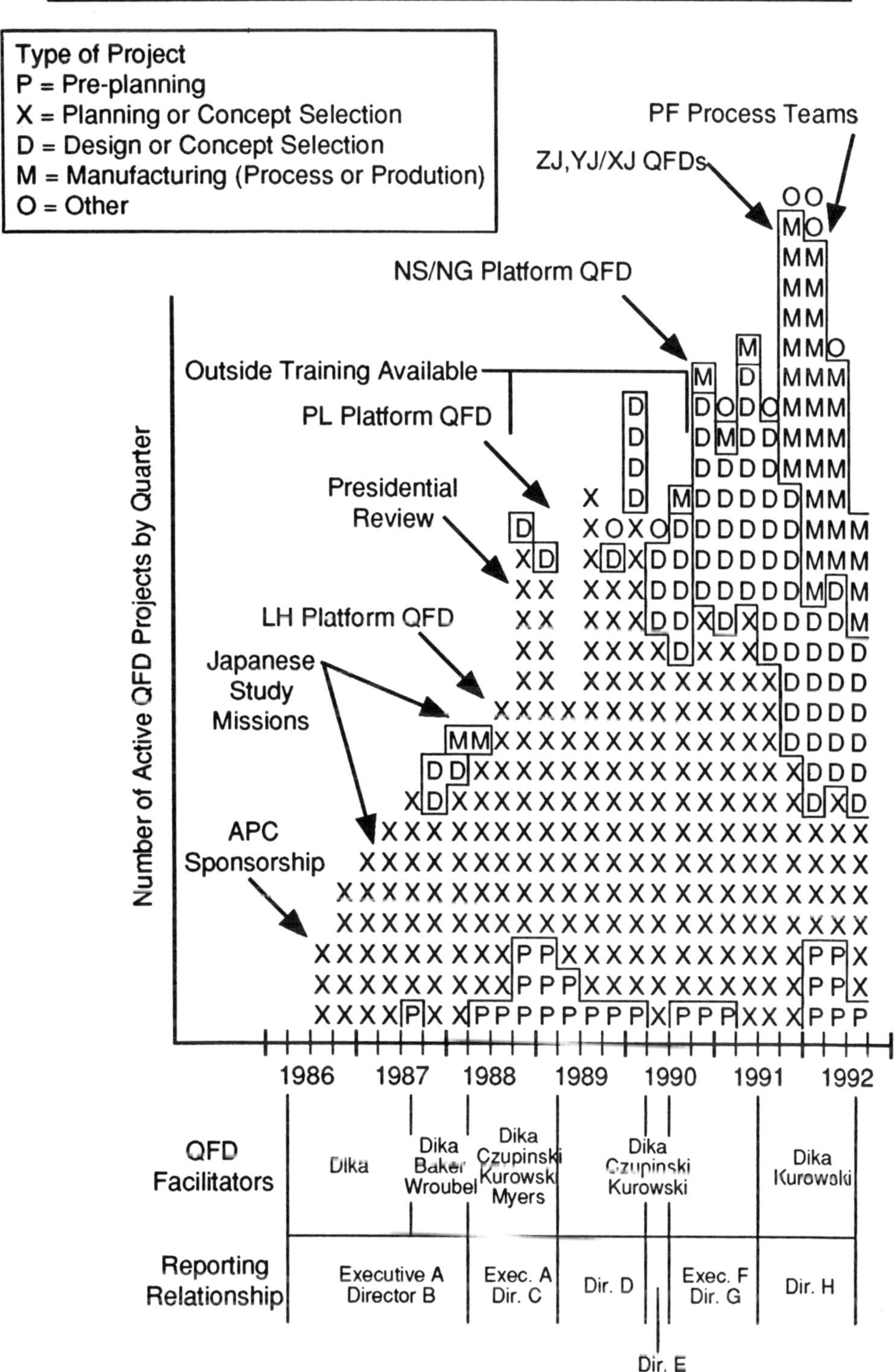
Type of Project
P = Pre-planning
X = Planning or Concept Selection
D = Design or Concept Selection
M = Manufacturing (Process or Prodution)
O = Other

PF Process Teams
ZJ,YJ/XJ QFDs
NS/NG Platform QFD
Outside Training Available
PL Platform QFD
Presidential Review
LH Platform QFD
Japanese Study Missions
APC Sponsorship

Number of Active QFD Projects by Quarter

1986   1987   1988   1989   1990   1991   1992

QFD Facilitators
Dika
Dika Baker Wroubel
Dika Czupinski Kurowski Myers
Dika Czupinski Kurowski
Dika Kurowski

Reporting Relationship
Executive A Director B
Exec. A Dir. C
Dir. D
Exec. F Dir. G
Dir. H
Dir. E

FIGURE 9–2
Approximate Portion of Overall Effort.

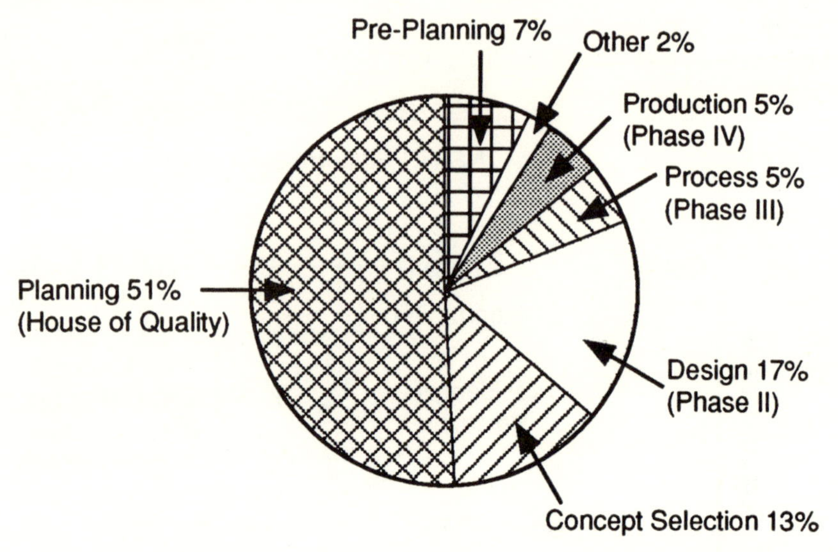

too easy. It turns out that not everyone sees QFD as the sensible approach that I do. Many people will always come from a show-me point of view. Some will never be convinced. In any case, implementation of QFD will not happen overnight.

On the plus side, we benefited from the use of what we have called the preplanning matrix for big projects, like a total vehicle, to help us establish overall priorities and to break down the task. We initially underestimated the attractiveness of the Pugh Concept and FMEA to teams working on product systems, but were able to use both to our benefit once we started to include them in our bag of tricks. We benefited by being able to maintain a small core of QFD fanatics who were able to stay together, learn together, and support people trying to pioneer in the use of QFD. And we benefited from the toughness and determination of many working-level people who became our team leaders and our most important supporters, even during the tough times.

On the negative side, we started without much support or standardized materials on the technique itself. We probably focused too much on the House of Quality and the first phase of QFD. We allowed early projects to take too long and to take side trips, like asking for market research before the team was mature enough to fully utilize it. We should have started by harnessing the collective experience of the teams and then moved more quickly to establish the entire QFD chain from planning to the shop floor. This is what we have done in recent years.

Almost all of the literature on QFD and Total Quality Management (TQM) says that it must start at the top or it won't work. In most cases, if you wait for it to start at the top, you will be waiting for a long time. In practice, change has to start where there is passion and there are people willing to drive for change. This spark can come from anywhere in the organizational hierarchy. In our case, top management endorsed what we were doing and allowed people at the working level to decide what worked and what didn't. They empowered their people and tried to let them do their jobs.

Senior people in any organization behave the way they do for good reason. They have built successful careers utilizing methods that have worked for them. Top management spends their time looking at the big picture, considering overall corporate performance, major turning points, and setting their vision and tone. They may view QFD as only a tool, a detail. We cannot expect them to change to our point of view instantly. They will change their behavior when they see that different behavior is more competitive. Our top managers didn't always understand what we were doing, but they were very helpful in speaking out supportively and often in asking the right questions at the right time. They saw our enthusiasm and trusted that what we were doing was valuable. I believe that QFD has been a positive influence in creating a change in culture that is now reflected by top management. I don't know if we could have made change any faster if implementing QFD was top management's idea in the first place.

We found that we had to be patient. Management people at all levels are busy. Those that are most critical to the growth of QFD are the engineers and middle managers directly responsible for producing the products and services of the company. These are the people that must become convinced if QFD is to succeed, and this is where we spent most of our time. Chrysler is now approaching what might be called TQM, although it did not happen through a deliberate Total Quality Management (TQM) program or directive. It happened from within.

We found that we tried to teach people too much too fast. Particularly in these times of downsizing, the people that really do the work are very busy. They are more receptive to new ideas in small bites. If you try to convince them to accept the whole scope of QFD at once, they cannot fathom it. As a colleague says, "If you show it all to them, they will throw up." Facilitators must develop a sense of how hard to push and wait for their time to speak up. If QFD can be offered as a tool to help people do what their boss has asked them to do, it is much easier to sell. The job of the facilitator is to be prepared to say the right thing at the right time with authority and to be as clear as possible as to the outcomes and the methods.

QFD does not work without a team. When we started, it was very difficult to get a team started. People at Chrysler had little or no experience working in teams. To an engineer, a crossfunctional team was a shift-level designer meeting with the shift-linkage designer. Early QFD teams not only had to learn QFD, but leaders had to learn how to form and lead a crossfunctional group. More often than not, it was difficult to convince people from manufacturing or marketing to attend. Today, teams are the norm at Chrysler. We have come a long way and it is much easier to proceed with QFD because most team members have decent team-building skills and expectations. I believe that our efforts to form and support QFD teams contributed to the current situation.

Finally, we can underestimate the value of informal applications of the principles of QFD. QFD can be applied at a variety of levels (see Figure 9–3). It can be used conceptually or informally, where people become aware of the importance of the customer and their requirements. They start to become involved crossfunctionally to talk about what the customer wants and needs. Results are not necessarily documented. This is QFD at the 100 level.

At the QFD 200 level, work is qualitative. Teams are formed, often in the normal course of business. Customers may be interviewed up-front and market research is conducted during the product-development cycle. Customer requirements are written down, but matrices may not be used. People start to report formally and set objectives based on customer satisfaction. Changes to product direction based on the information are real and substantive. QFD may help to

FIGURE 9–3
Learning Curve.

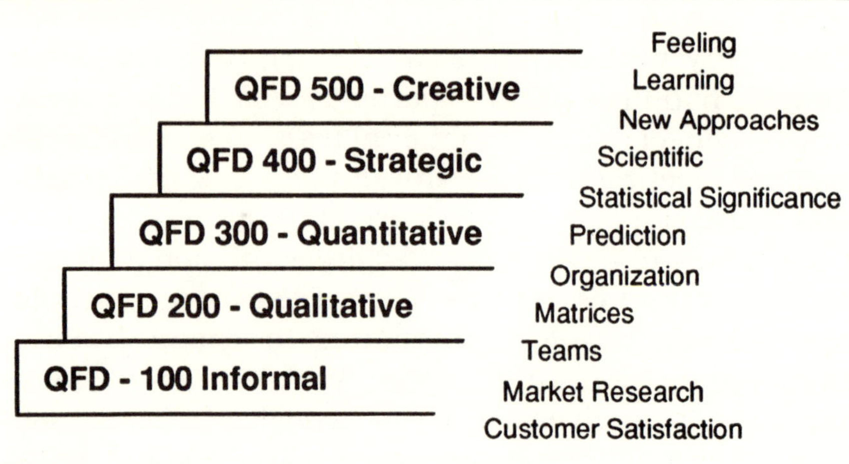

focus reliability and process capability work on the critical areas. It may be used as a way to influence the culture or ''soft-side'' of the business.

At the QFD 300 level, work begins to be quantitative. Teams are organized for success. They have a real mandate from management and good cross-functional participation. Formal meetings are held and the teams construct matrices that are distributed and utilized to facilitate decision making and communication. QFD may start to support a TQM system. This is what most people think of as QFD.

At the QFD 400 level, QFD becomes a scientific, strategic weapon. It becomes a technological tool to fight for market share. Research and the science of QFD becomes the focus. Statistically significant data and logic algorithms are used to support decision making. QFD is used to identify critical areas and other quality and reliability tools are used to solve the problems identified by the QFD projects. Process capability becomes a truly meaningful metric. Results are correlated to planned levels of performance. At the 500 (post-graduate) level, QFD is used to facilitate creative learning. Teams become interested in expanding the QFD envelope and discover new ways to use it. They want to discover how to create an intangible feeling of enduring delight for the customer.

At Chrysler, we have not always documented our QFD studies as well as we would have liked, so it becomes hard to discuss the effects that projects have had. Since there is no comparable control, there is no way to show that QFD caused the team to have performance greater than if QFD was not used at all. But I believe that many of the soft applications of QFD, at the 100 or 200 level, had significant influence on both the product and the culture. QFD can have many intangible benefits, with no way to demonstrate them. These applications can have powerful, transformative effects on an organization, but go unnoticed by those trained to believe that progress must be measured only in dollars and cents.

Some of us have grown tired of pundits and critics who try to study the cause-and-effect relationship between the use of QFD and corporate performance. They are not helping. What is needed are people willing to help make it work.

YEAR SEVEN

Year seven—1993—will be a year of transition for Chrysler. In December 1992, the chairman retired. The new management team will undoubtedly start to establish itself, and the company may find itself taking a new direction. Although there is still a long way to go, the hard work of the last few years has finally put the company in a position to introduce new, attractive, high-quality products steadily over the next few years. Consumers are already starting to respond. With the strong showing in the fourth quarter of last year, it is felt that the economy is finally making a solid upturn, and the car business will be good.

At the end of its seventh year at Chrysler, QFD will also be going through a transition. The essential message of this paper is that small teams, working hard, can make a difference. QFD has grown in both the number of projects and in the significance of those projects on both the product and the people. It has helped to improve the company's fortunes and the development of a more customer-driven, team-based culture.

Teams and teamwork are now the standard mechanism for conducting business. QFD has been well inoculated into the product-assurance system. The elements of QFD are officially recognized in the work plans of teams at the vehicle, system, and component level in all of the platforms. Training in QFD is again available. Many people with experience and belief are spread throughout the organizations.

There is no longer a centralized QFD group in engineering or in corporate staff. People with facilitation experience are now located in design engineering, vehicle development, quality planning, and in other functional organizations. The next stage of QFD growth at Chrysler will have to spring from within the line organizations and platforms.

The pump has been primed and I am confident that QFD will continue to grow. The following anecdote might shed some light on why.

On a bright day in February, a colleague of mine was walking down the hall and was stopped by the general manager of the small-car platform, with whom he had not exchanged more than a few words in some time. This general manager had trusted his people and QFD work teams in the face of persistent rumors that QFD was not totally supported by top management. He was beaming and anxious to talk, knowing that my friend was a QFD advocate. He had just returned from a week-end customer research event in San Diego, where the PL, now the new Dodge and Plymouth Neon, still a year from introduction, had been evaluated.

Not only did the customers like the Neon, but they were overflowing with praise. They said that they could imagine a young married couple, on vacation, in the PL with the windows down having fun driving up the California coast. In their explanation of what they liked about the car, the research participants kept men-tioning terms like "well made," "value," "fun to drive," and "roomy," the very attributes that our QFD project identified as critical three years before.

Following the research, the cars were shown to a group of writers at the Chrysler Arizona Proving Grounds just outside of Phoenix. The executive engi-neer of vehicle development made a presentation on the QFD study. He talked about how we identified up-front what we wanted to do and how we kept measuring our progress against the customer's requirements, particu-larly in the area of "fun-to-drive." The press people were similarly excited, both about the Neon and the changes that we had made in our development process.

The general manager then, with a look of delight, took my friend into his confidence. "You know, this s _ _ _ really works!" he said.

ACKNOWLEDGMENTS

This story would not have been possible without the individual contributions of hundreds of people. The author would especially like to acknowledge the following principal activists in introducing QFD at Chrysler.

Alan C. Carlson, executive engineer, small-car platform program management.

Chris W. Kurowski, specialist, quality Synthesis, small-car platform vehicle development.

Glenn W. Czupinski, specialist, minivan platform body and interior systems engineering.

Ray L. Begley, specialist, quality Synthesis, small-car platform vehicle development.

Monte G. Myers, manager, quality Synthesis, small-car platform vehicle development.

Henry J. Ziaja, specialist, quality Synthesis, small-car platform vehicle development.

REFERENCES

Akao, Yoji. *Quality Function Deployment—Integrating Customer Requirements into Product Design,* 1990.

American Supplier Institute. *QFD Three-Day Workship Implementation Manual,* Allen Park, MI, 1987.

Dika, Robert J., and Ray L. Begley. "Concept Development through Teamwork—Working for Quality, Cost, Weight, and Investment." SAE Technical Paper Series No. 910212. SAE Congress, February 26, 1991.

______, et al. "Process Improvement in Chrysler's Small-Car Platform." SAE Technical Paper Series No. 930471. SAE Congress, March 2, 1993.

Eureka, William E., and Nancy E. Ryan. *The Customer-Driven Company—Managerial Perspectives on QFD,* 1994, Irwin Professional Publishing and ASI Press.

Ford, Henry. *Today and Tomorrow*, 1926.

Hauser, John R., and Don Clausing. "The House of Quality." *Harvard Business Review*, May–June, 1988.

Ishikawa, Laoru. *Guide to Quality Control*, 1992.

______. *What Is Total Quality Control? The Japanese Way*, 1985.

King, Bob. *Better Designs in Half the Time—Implementing QFD in America*, G.O.A.L., Inc, Lawrence, MA, 1987.

Monden, Yasuhiro. *Toyota Production System*, 1983.

Ohio, Taiichi. *Toyota Production System—Beyond Large-Scale Production*, Productivity Press, Cambridge, MA, 1988.

Report to Shareholders. Chrysler Corporation, Highland Park, MI, 1986–1992.

Shingo, Shigeo. *Zero Quality Control: Source Inspection and the Poka-yoke System*, 1986.

Sullivan, L. P. "Quality Function Deployment." *Quality Progress*, Milwaukee, WI, June 1986.

Transactions from "A Symposium on Quality Function Deployment," American Supplier Institute, Allen Park, MI, 1989–1992.

Ziaja, Henry J. "Total Product Quality Process Model." ASQC 44th Annual Quality Congress, Milwaukee, WI, May 15, 1990.

Chapter Ten

Implementing Simultaneous Engineering: Achieving Robust Designs Based on the Bottle Model

William E. Eureka

Robert Moesta

BACKGROUND

As the global economy becomes increasingly competitive, improving our effectiveness in bringing superior products to market is essential. By superior products we mean high-quality, low-cost products that can be brought to market very quickly.

William E. Eureka is a learning coach at Herman Miller Inc., Zeeland, Michigan.

Robert Moesta is a principal of Pedi, Moesta & Associates, a consulting firm in Grosse Pointe Farms, Michigan.

FIGURE 10–1

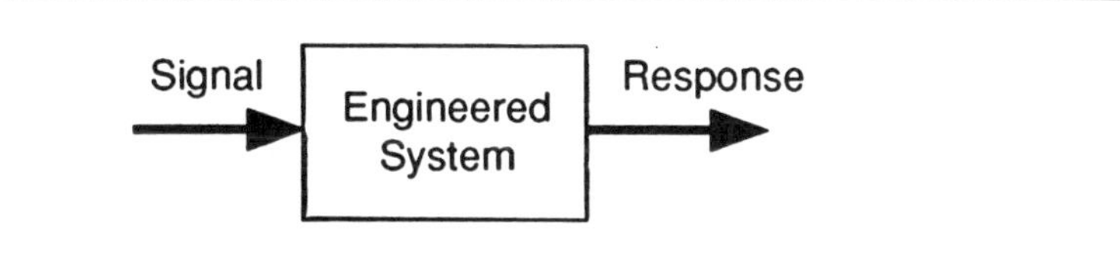

Furthermore, these products must be fully developed when they reach the market. The customers will no longer tolerate being our lab testers.

This requires that we increase the product's robustness. We can consider robustness as the ability of a product to perform as expected, even when faced with forces or conditions that tend to degrade its performance.

Thus, a robust product would perform well in the laboratory as well as in diverse usage conditions—even some customer abuse. More robust products will perform more consistently than less robust products. The challenge is to engineer robust products quickly and at low cost.

THE ENGINEERED SYSTEM

We may think of anything we wish to design (no matter how simple or complex) as an engineered system in which some signal (an input) is acted upon to produce a response (an output), as shown in Figure 10–1. All products and processes may be viewed in this way.

In order for the system to be useful, it must work. While this may appear obvious, considerable R&D effort is expended in developing technologies that will work.

FIGURE 10–2

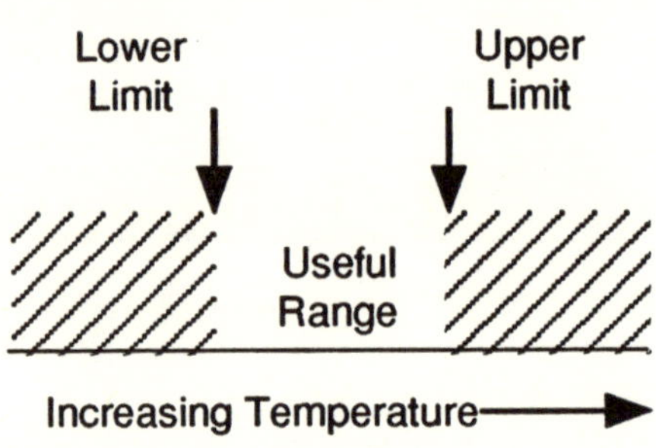

When workable technologies are developed, they often function over a limited range of operating conditions (see Figure 10–2). Such technologies are highly sensitive and not very robust. Products that use such technologies must be nursed through lab testing and will often have many unexpected failures in actual usage.

If we think of the many variables that describe the technology's design parameters and operating environment (such as ambient temperature), we find that there is typically a range within which the technology is usable. Outside of that range, the system will either fail, operate at an unacceptable performance level, or exhibit undesired side effects.

We can think of the useful range of operation as the operating window of the system. A system with a closed operating window simply will not work. Those with open operating windows will work. Immature technologies often have an operating window that is open just a crack. Our product development process focuses on how to open the operating window as much as possible,

FIGURE 10–3

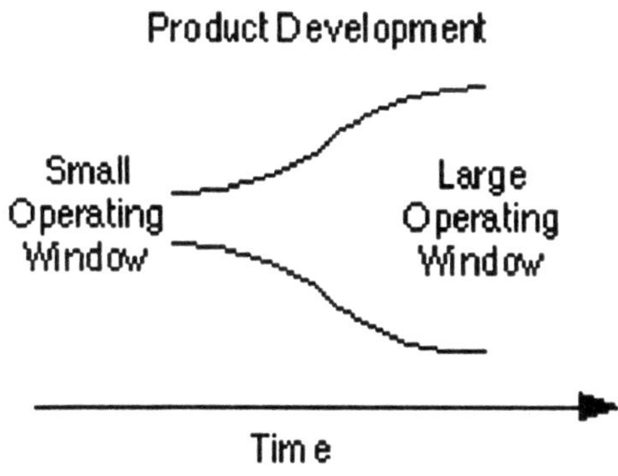

quickly, and at low cost. We will now focus on how this is currently done, then on how to do it better.

Figure 10–3, which shows increasing design latitude over time, is called the *bottle model.* Our normal product development system opens the system operating window, using some of the following approaches:

- Peer design reviews.
- Design Failure-Mode and Effect Analysis (FMEA).
- Product testing.
- Simulation and math modeling.
- Design of experiments.
- Quality Function Deployment (QFD).
- Failure analysis of prior products.

During the implementation of simultaneous engineering (also known as concurrent engineering or integrated product development), increasing emphasis is being placed on ensuring compatibility between product and process design. This includes ensuring that the product design can be manufactured. Typically, process capability is compared to the product tolerances.

If process capability is inadequate, either design or process changes are made to achieve compatibility. As long as the manufacturing variability is well within the tolerance range, we are often comfortable that the product and the process are compatible. Unfortunately, this will only ensure that the product can be *manufactured* within tolerance—not that the product will *work* under a wide range of conditions.

Robustness was described earlier as the ability of a product to perform as expected, even when faced with forces or conditions that tend to degrade its performance. These "forces or conditions" are called noise factors, and include any issues that influence the design, but are beyond our control. Noise factors can also include issues that we deliberately *choose* not to control for cost, technological, or other reasons.

Noise factors may be grouped into three broad categories:

- Product-to-product variation.
- Product degradation over time.
- Variation in customer usage and environments.

We can address product-to-product variation by improving our manufacturing process capability, but we cannot greatly influence the other types of noise factors in our factories. These other types of noise factors are the responsibility of product design.

Because there are many noise factors beyond the realm of manufacturing, our products should have the widest possible design latitude (or very wide operating windows). A wide design latitude will ensure that our

FIGURE 10–4

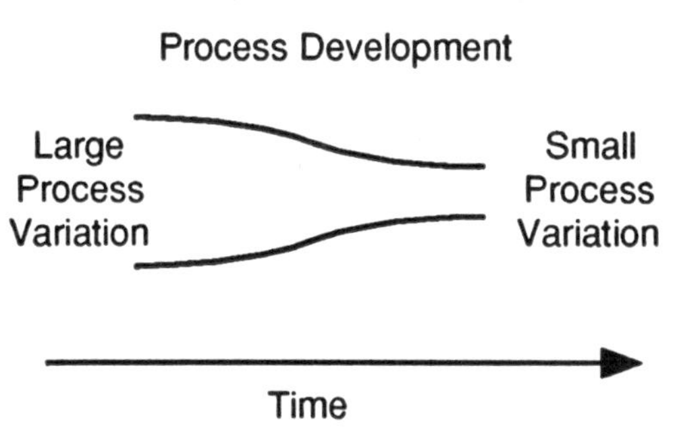

products will operate more consistently with customer use.

When designs are deemed producible because their tolerances are wider than the process variation, there still may be insufficient design latitude to accommodate the noise factors that will be found with customer use.

While increasing the design latitude is pursued in product development, our process development activities should continually seek to reduce manufacturing variability.

The process development effort follows its own bottle model, necking down over time, as shown in Figure 10–4. This results in improved process capability as the process develops *before* production launch.

Our normal process development efforts help reduce variation using some of the following approaches:

- Process reviews
- Process FMEA
- Process testing
- Design of experiments

FIGURE 10–5

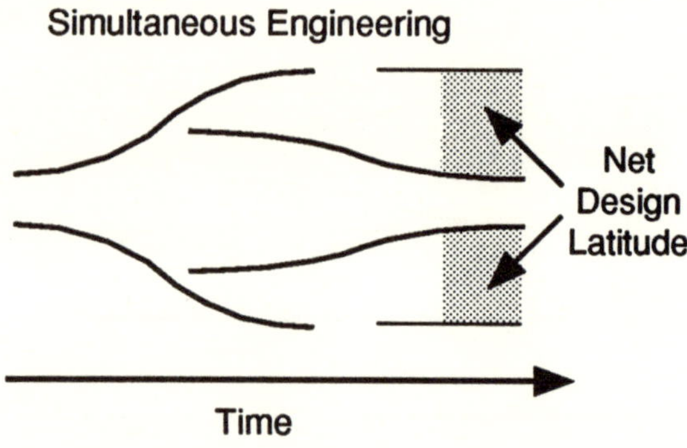

- Process control plans
- Prior problem-solving efforts.

If we overlay the product and process bottle models, we see that they work together to achieve a better product.

By opening product design latitudes, while reducing process design latitudes, we increase the net design latitude, or the amount of design latitude available to meet the customer's needs (see Figure 10–5). The greater the net design latitude of the product or process delivery system, the greater its robustness.

IMPROVING DEVELOPMENT

While the normal product and process development systems of successful companies do improve net design latitude, in order to gain further competitive advantage, they must become even more effective. The key answers needed include:

FIGURE 10–6A

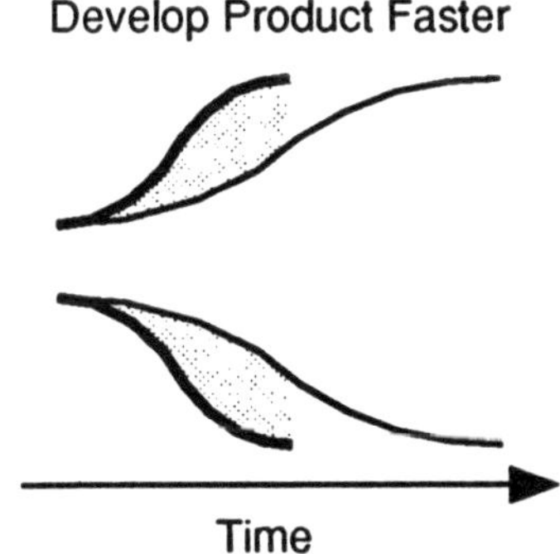

- How to increase design latitude faster.
- How to reduce process variation faster.
- How to begin development using technology with greater design latitude.

Increasing design latitude faster means widening the operating window faster. This allows for more rapid development of the product design, as shown in Figure 10–6A.

Reducing process variation faster means more rapidly discovering and acting upon the causes of variation. This allows for more rapid evolution of the process design, resulting in quicker production readiness (see Figure 10–6B).

Developing technology with a greater design latitude allows that technology to evolve *before* introducing it into a new product (see Figure 10–6C). This reduces program risk while dramatically reducing overall development time.

FIGURE 10–6B

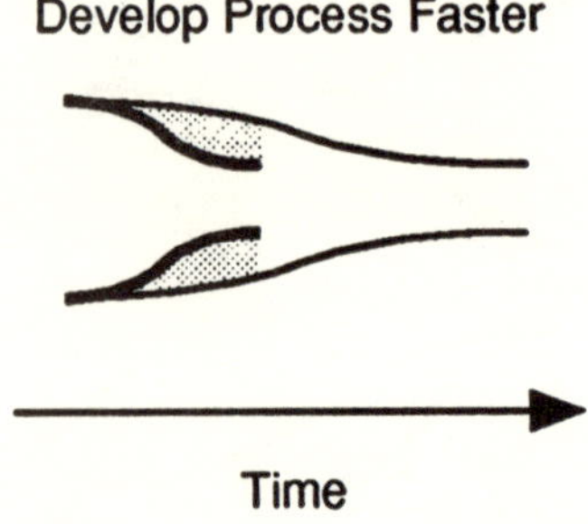

FIGURE 10–6C

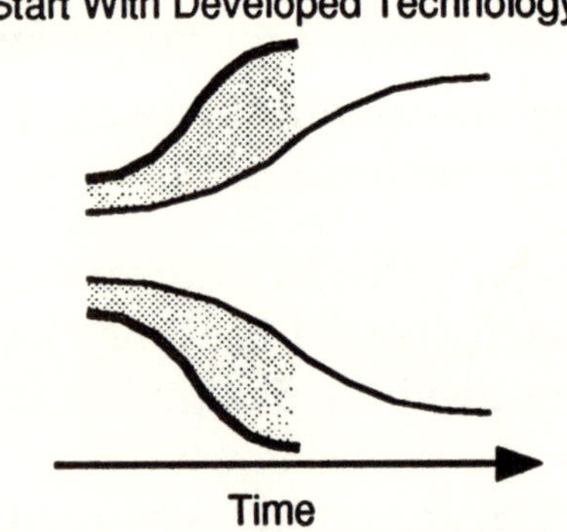

Making substantial improvements of this type requires a fundamental change in our engineering approach—the very paradigms that guide our engineering thinking.

OUR ENGINEERING PARADIGMS

Our paradigms provide us guidance in the form of models, or "rules of thumb," that allow us to quickly make correct decisions without having to carefully study each situation.

This is useful so long as our paradigms fit the situation. If they don't fit, they will steer us to the wrong actions. Some current paradigms that need to be seriously questioned include:

- Fixing problems improves the design.
- Improving quality takes more money and time.
- Prototypes are for debugging the design.
- It's only a prototype; we'll be closer on the next iteration.
- We can't make further improvements until we have hardware.
- We can't test that system until it's built into a completed product.

These paradigms, like many others, may have served us well in the past, but can now become ballast weighing us down.

The paradigm shown in Figure 10–7 is one of the most important to change because it is central to the way we develop products.

This engineering paradigm begins with our creative abilities to devise a workable design (with on open operating window). We see problems as the obstacles that block our design efforts. By finding and fixing those problems through design revision, we improve the design.

FIGURE 10–7

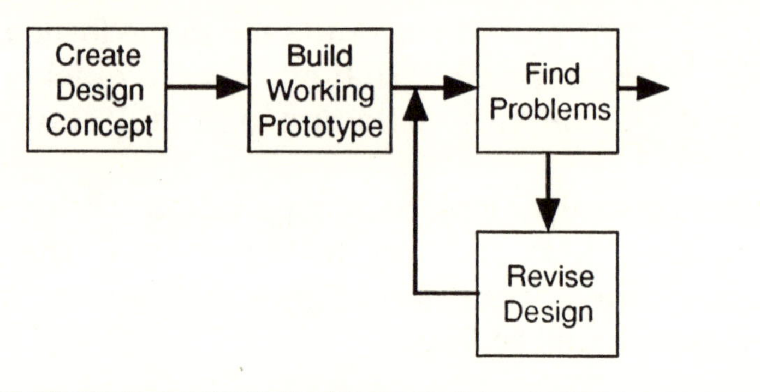

While this approach usually does work, it is far too slow to aggressively compete in today's market. It is based on achieving success by removing the causes of failure. Unfortunately, when we look for problems, we usually find them. In fact, there never seems to be an end to them!

A NEW ENGINEERING PARADIGM

A more effective paradigm focuses on achieving success by improving *things gone right,* rather than by reducing *things gone wrong.* To do this we focus on improving the *function* of the system. The function defines the relationship between the input signal and the output response.

The ideal function of the system defines the energy transformation that converts input energy into useful response energy. This ideal function is linear because it maps the customer intent (the signal) to the desired system result (the response), as shown in Figure 10–8. This

FIGURE 10–8

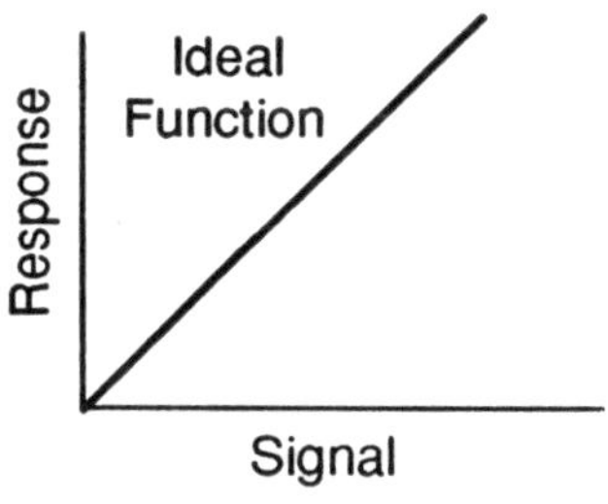

ideal function assumes perfect energy transfer without loss. Since this can never occur because of energy loss, the actual function will be less than the ideal function (see Figure 10–9).

Where does this lost energy go? Since energy is neither created nor destroyed (except in nuclear reactions), it manifests itself in undesired forms, such as heat, noise, fatigue, vibration, and reduced life.

All of these are forms of problems—all caused by energy in the wrong place.

When we focus on solving a specific problem, we attempt to reduce the energy causing that problem. In doing so, we often inadvertently redirect that energy into yet another problem. The net result is one of "chasing" problem after problem while debugging the design.

The new engineering paradigm shown in Figure 10–10 focuses on maximizing the proportion of response energy that is useful, leaving very little energy to create problems. This also causes the actual function to be closer to the ideal function.

Maximizing this proportion improves robustness, or design latitude. This energy transfer model provides a

FIGURE 10–9

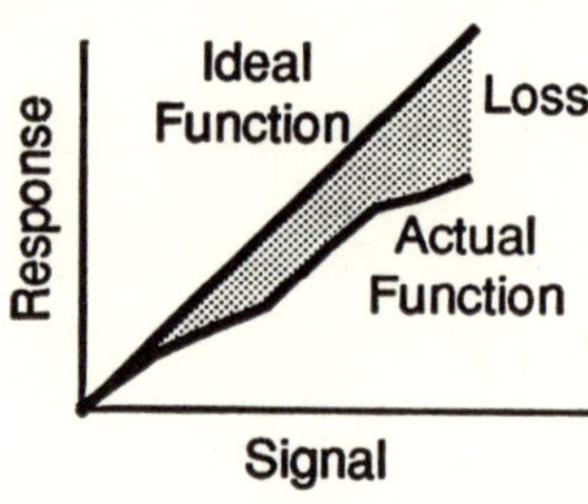

measure for robustness, called the Signal-to-Noise (S/N) Ratio:

$$S/N = \frac{E_{useful}}{E_{waste}}$$

By maximizing the S/N Ratio, we maximize the robustness by channeling the most energy possible into the useful function with less wasted.

ACCENTUATING THE POSITIVE

To improve functionality, we need to understand why energy is diverted from the ideal function into problems.

Problems happen when we work outside the operating window. As we widen the operating window, fewer problems remain and the actual function approaches the ideal function.

The width of the operating window is set by:

- Design concept chosen.
- Values of control factors.
- Noise factors.

FIGURE 10–10

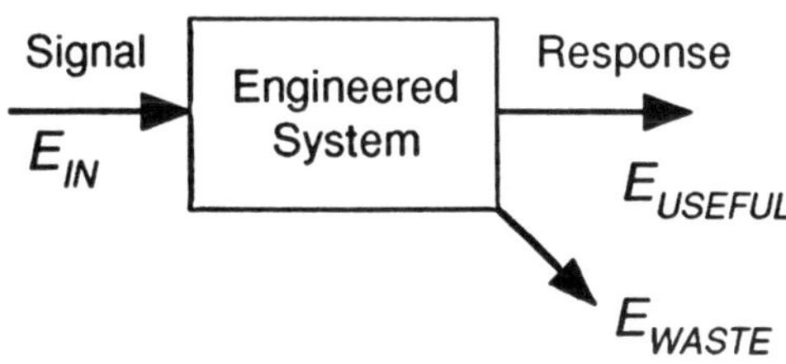

Some design concepts inherently have more design latitude than others. To improve this latitude, we will seek to choose the best design concept possible using known and well-developed technology. Control factors are the design parameters that we consciously set to make the design function correctly.

The three types of noise factors (product-to-product variation, product degradation over time, and customer usage and environments) are beyond our control. There are four approaches typically employed to deal with noise factors:

- Cause removal
- Compensation
- Doing nothing
- Robust design

Cause removal sounds attractive because it gets rid of the problem at its source. Unfortunately, many noise factors simply cannot—or are too expensive to—be removed.

Compensation is the most frequently used counter-measure to noise. Compensation accepts the presence of noise factors and seeks to neutralize them by adding something. Examples of compensation include:

- Feedback systems
- Feedforward systems
- Redundancy
- Overdesign

Compensation is seductive because the result can be an elegant engineering solution that triumphs over the problem. Unfortunately, it almost always adds cost and complexity, leading to higher product and manufacturing costs and lower reliability.

Doing nothing in response to noise factors is the approach most often taken when technology is insufficiently developed. When this is the case, there is not enough experience to anticipate potential problems.

The result of doing nothing is that problems will eventually become evident late in the design cycle. The "find and fix" approach is then applied, increasing cost and time, until a solution is found. This is one of the best reasons to avoid using undeveloped technology on a new product design.

Cause removal and compensation are reactive approaches used after experiencing problems (either in the current design or prior similar designs). Doing nothing becomes a reactive approach when problems are finally discovered.

All three approaches focus on reducing things gone wrong and require prior knowledge of what those problems might be. FMEA, Fault Tree Analysis (FTA), warranty analysis, and customer complaint analysis are approaches commonly used to identify such potential problems.

Such approaches can never be complete without a great deal of experience with the design concept and its technology. We face the dilemma of new technology with a multitude of potential problems that are beyond our current experience.

The fourth approach to noise, robust design, is the only one that can be applied effectively when our knowledge of potential problems is immature. This is because it focuses not on problems (which must be known or suspected before they can be addressed), but on optimizing the ideal function of the engineered system.

Through robustness, we increase the S/N Ratio to maximize the energy transfer into the useful function, thereby minimizing the wasted energy available to create problems. In this way we prevent problems without explicitly knowing them. This approach is enormously more efficient than chasing after the many potential problems that could plague a complex design.

ACHIEVING ROBUST DESIGN

While the robust design approach is the most powerful way to maximize design latitude, it is the least understood and least often used. To understand this approach, it is necessary to discuss another aspect of robustness. So far we have stated that robust products or processes have the following characteristics:

- Minimum sensitivity to forces or conditions that tend to degrade its performance.
- Wide operating window (maximum design latitude).
- Actual function that approaches the ideal function.
- Very few problems.
- Maximum S/N Ratio.

FIGURE 10–11

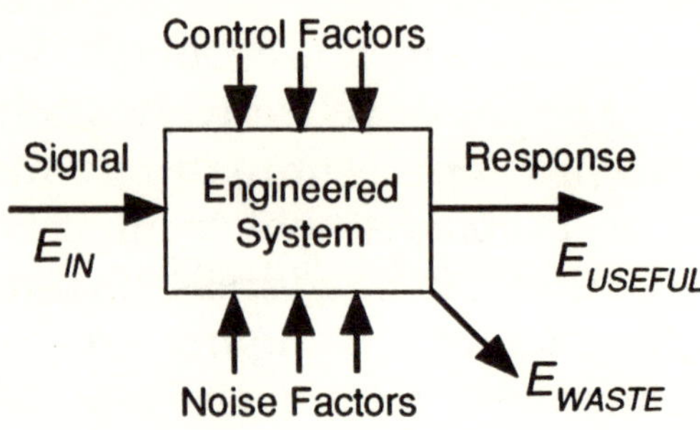

Furthermore, robustness is achieved by the particular way noise factors are addressed. The robust design approach accepts the noise factors, but it seeks to find the best values of the control factors that will minimize the influence of the noise factors, as shown in Figure 10–11.

The robustness approach begins by selecting either the response of the ideal function or a more convenient surrogate. This surrogate, sometimes called the quality characteristic, is chosen when the ideal function is difficult or costly to measure directly.

The choice of the proper quality characteristic is critical! A wrong choice may be a waste of time and resources at best and degrade the design at worst. Because we are attempting to optimize an ideal function that is an energy transformation, the quality characteristic should be closely related to the response energy.

Next, all factors that are believed to influence the quality characteristic are listed. This is best done using a knowledgeable group of people. Knowledge is the key here. If this activity is done by a product development team, be sure to invite others whose participation or suggestions may be useful.

The factors are then grouped into control factors and noise factors. Control factors are the design parameters that we consciously set to make the design function correctly.

For products, control factors might include:

- Design configuration alternatives.
- Drawing specifications.
- Material alternatives.
- Material specifications.

For processes, control factors might include:

- Process alternatives.
- Settings of process controls.
- Operating methods.

Control factors are within *our* control. The speed at which a customer operates a product is not a control factor. Although it is a factor that is controllable, it is controlled by the customer, not by our engineering efforts.

Noise factors are typically beyond our control. They might include factors that are:

- Impossible to control.
- Difficult to control.
- Too costly to control.
- Undesirable to control.

While the last two items may be technically controllable, we choose not to control them as a pragmatic business decision.

The philosophy behind the robustness approach was described by Reverend Reinhold Niebuhr, who in the 1950s penned a verse that he called his Serenity Prayer:

> God grant me the serenity to accept
> the things I cannot change,
> The strength to change
> the things I can
> And the wisdom
> to know the difference.

While it is unlikely that Niebuhr knew much about engineering, his philosophy makes good engineering and business sense. Work on what you can and don't let the other things give you heartburn!

We apply this principle in robust design by choosing the values of the control factors that will minimize the *influence* of the noise factors.

Thus, we control the things we can in a way that minimizes the effect of the things we cannot. This approach to design was first formalized by Dr Genichi Taguchi. He called it parameter design. Parameter design is used to determine the best (functional and economical) target values for the control factors that minimize the influence of noise factors. Taguchi also devised an approach called tolerance design, which determines the proper balance between the tightening of tolerances and cost implications.

To do this, we need knowledge of how the control factors and noise factors work together. If we do not already have such knowledge, it is normally discovered through analysis, simulation, or physical testing.

As shown in Figure 10–12, the choice of the control factor value affects the result (the value of the quality characteristic). This is expected, since it is a control factor. What is often unexpected is that different control

FIGURE 10–12

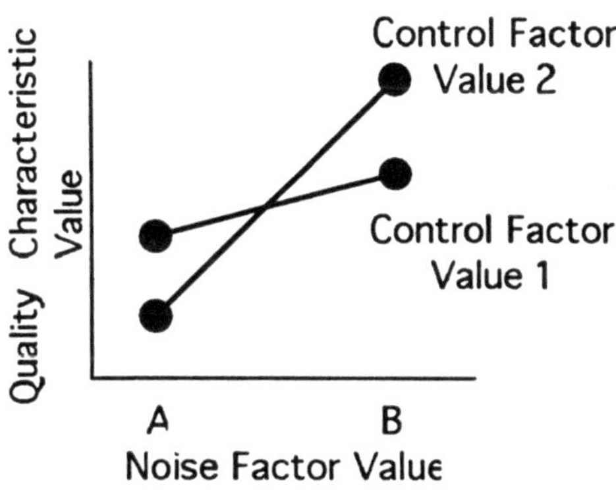

factor values result in different levels of sensitivity to noise factors.

When the above control factor is set at its second value, the maximum value for the quality characteristic may be higher, but the minimum value is also lower. The actual result we get is out of our control, subject to the whims of the noise factors. The first setting is much more stable and insensitive to noise.

If we choose the wrong values for the control factors, we may get good lab test results only to be surprised by problems in the hands of customers (where noise factors abound).

As we explore the relationship between control factors, noise factors, and the quality characteristic, we can usually discover substantial cost-reduction opportunities. Many designs have a large number of control factors. It is inconceivable that they would all have the same influence on the result. There is always some difference in their influence. As we discover that some of the control factors have a large influence, we will set them to achieve the stable performance level desired. Less influential control factors can be set at their lowest-cost value. This reduces product and process costs early in the development cycle, helping to preclude the need for disruptive, crash cost-reduction efforts late in the cycle.

DEVELOPING PRODUCTS FASTER

Products that evolve faster can be implemented more quickly. The build-test-fix approach that specializes in problem solving is much too slow. Robust design will help widen design latitudes faster (see Figure 10–6A).

Robust design techniques are used to develop new products faster through the following steps:

- Define the ideal function.
- Explore feasible alternatives.
- Select the best alternative.
- Optimize the best alternative.
- Build and confirm.

Function.   The ideal function should be an energy-related transformation of signal into response at the lowest level of detail. When this is done, the ideal function will tend to be linear because energy is additive.

A focus on nonenergy related functions will lead us astray by concentrating on symptoms or side effects (many of which are highly interactive). Focusing at too high a level will lead to functional relationships too complex to understand.

Alternatives.   Exploration of alternatives seeks to investigate the range of potentially feasible approaches to achieve the ideal function. It is important to look beyond traditional approaches, to seek better ways. Remember that the purpose of the design process is not to design something *new* (anyone can do that)—it is to design something *better!* It is important to have multiple alternatives because the first idea we have is almost never the best idea. More ideas give us a better chance for improvement.

We all have our paradigms of how the systems we design are created. It's how we do things in our organization. While this approach may work well, it tends to lead to more similarity than innovation. When we benchmark our competitors' products, we begin to step out of our paradigms and into theirs. While we may look upon our competitors' products with disdain, there is usually something to learn.

Benchmarking noncompetitive products may provide truly original insights not considered by our competitors. Noncompetitive companies may also be willing to share technical information that competitors would not. One manufacturer of automotive paint spray guns studied the technology used by ink-jet computer printers. While the technology and its application are very different from that used in painting cars, the paint spray gun manufacturer was able to discover useful ideas from the products.

Value Engineering (VE) is also important at this time. VE seeks to maximize the value of the product, defining value as:

$$\text{Value} = \frac{\text{Function}}{\text{Cost}}$$

While VE was introduced in 1947 by Larry Miles of General Electric, it is often overlooked because it is believed to be too much work for the results. The price of *not* using VE often is unnecessarily complex designs. Complexity is remarkably easy—add something to a prior design and it becomes more complex. When we use compensation, we usually add complexity to the design. Simplicity can be extremely difficult to achieve.

Based on the simple VE formula, we can improve value by improving function, reducing cost, or both. Many VE efforts focus on reducing cost with little or no degradation of function. While this is useful, it is more productive to take a balanced approach early in design, since function and cost are almost always interdependent.

By focusing on function, we concentrate on what we *want* to achieve rather than what we don't want (compensation, problems, wasted energy). Considering cost as well will help us select an alternative with the best commercial prospects.

In *The New Experimental Design*, Teruo Mori discusses value (which he calls utility) and asserts that "cost is dependent . . . on the technological capability of the corporation's engineers."

Design for Assembly (DFA) can also contribute at this point. DFA was developed in 1980 by Geoffrey Boothroyd and Peter Dewhurst of the University of Massachusetts. Like most improvement approaches, DFA is most influential when applied as early as possible.

DFA is often used to redesign a system so that product and manufacturing costs can be reduced. Redesigning a system that is already in production may be beneficial, but considerable investment has already been made in the current design and processing. Many are using DFA to redesign before prototype, avoiding costly investment in tooling, equipment, and facilities.

While some of the methods of DFA cannot be applied before a reasonably detailed design is available, the philosophy of DFA is useful even in determining design concept alternatives. These include (1) reduction of part count and (2) reduction of time to assemble parts.

Benchmarking, VE, and DFA can be combined and used with our own creativity to help generate a wide range of possible design alternatives.

Selection. Once a number of design concept alternatives are available, we must choose one for our product. While each competing concept has good and bad points, choosing the one with the best combination is the objective. This requires that we compare optimized concepts and compare concepts objectively.

Optimizing concepts may be difficult at this point, but we should take any feasible steps to avoid making all the concepts robust before selection. If we don't, we can easily choose an inferior alternative over a better one that is simply not totally optimized. While physical mock-ups can be used at this stage, mathematical modeling is preferred wherever possible.

It is important for everyone involved in the selection process to dispassionately compare the alternatives. This means not "falling in love" with your idea. As good as your idea may be, it can usually be improved. If one of the alternatives is your idea, try not to focus on it. Rather, probe into how it could be improved.

Objective comparison requires that there be a basis for comparison. This is accomplished by identifying selection criteria, by which we evaluate the merit of each alternative. The selection criteria should always include functional requirements, global specifications derived from customer wants, and any other important business criteria (always including cost).

While many comparison approaches exist, one of the most useful is Pugh Concept Selection, developed by the British designer Stuart Pugh. This approach focuses on comparing concept alternatives in a way that leads to the creation of new concepts that are better than the ones being compared. By using simple comparisons like "clearly better," "clearly worse," or "about the same," we focus first on improvement rather than choosing the "best."

Optimizing.   Once the best concept is chosen, it is detailed and optimized using the robust design approach, maximizing the S/N Ratio. This means determining the control factors and noise factors that influence the ideal function. Optimization involves determining the *values* of the control factors that maximize the energy transfer to useful function. This results in minimizing the effect of the many noise factors outside our control, as shown in Figure 10–10.

Both physical testing and mathematical modeling are appropriate at this time. Modeling is preferred to quickly converge on the most promising values. This may be confirmed through physical testing.

Confirmation.   The design is then built and tested to confirm its practicality. Note that we are not seeking weaknesses or problems because we do not expect surprises at this point. Our earlier optimization efforts before and after selection should have identified those.

Confirmation testing should include the following:

- Simulation of actual usage conditions.
- Measurable (numerical) results.

Care should be taken to carefully evaluate the test method. Many companies have time-honored tests that they feel obliged to use. As technology is rapidly changing, every test should be regularly challenged for continued usefulness. If the test does not represent actual usage conditions, what does it represent? How will we use those results?

Measurable results are easy to overlook. Many tests are attribute tests, producing pass/fail results. Such tests may subject the product to an ordeal; if it survives, it passed; otherwise, it failed.

This approach, which hopes to find problems if they exist, is the opposite of what we want. If the test is passed, we conclude (hopefully correctly) that the design is free of the problems the test should find. In this situation, we could have substantially overdesigned the product and not know it.

When problems are found, we often agonize over the unwanted surprise. Sometimes the test is condemned as unrealistic or too severe. If the problems are real but "rationalized away," they can be propagated through the design process right into the customer's hands.

If the problems are confronted, the design must be revised and retested, slowing down the devclopment process (see Figure 10–7). By focusing on measurable results, we can relate to earlier optimization results, learning more about the design while improving our optimization process.

DEVELOPING PROCESSES FASTER

Reducing process variability (improving process capability) faster allows rapid implementation of the new dcsign. As in product design, the robust design approach helps us get there faster, as noted in Figure 10–6B.

The rapid improvement of process capability can be done in much the same way as product improvement.

All the issues discussed for a product apply to development of new processes:

- Define the ideal function.
- Explore feasible alternatives.
- Select the best alternative.
- Optimize the best alternative.
- Build and confirm.

One difference is that processes often already exist. To minimize capital investment, we often make the business decision to use existing equipment and facilities instead of developing new ones. This provides tremendous opportunities because we already have both experience and physical hardware to optimize.

Robust design can be applied to existing processes by first defining the ideal function of the process and then optimizing the control factors to minimize the impact of noise. Control factors will include the "dials and knobs" on the process that we adjust to make it work properly. Noise factors will include all the other issues that influence the process.

Process optimization increases the design latitude of the process (allowing it to work well despite making a wider variety of products), while reducing the product-to-product variability.

Process optimization can also provide a proprietary advantage. Competitors may be able to obtain your products and "reverse engineer" them, but your processes are more secure. The more you understand your processes, the more you can push them to perform bet-tcr. Some companies know much more about their processes than the builders of the equipment. That type of competitive advantage is hard to steal!

DEVELOPING TECHNOLOGY FASTER

The rapid development of technology is an opportunity few companies fully pursue. As shown in Figure 10–6C, the potential to reduce development time and rapidly capitalize on short-lived technological advantages is tremendous.

It is natural to want to include the latest technology in our products. When our technology is only partially developed, there are many undiscovered and unresolved problems. In some cases, an attractive new technology may have an operating window that is barely open.

When we include undeveloped technologies in new designs, we greatly increase our risk of program delays, no-build problems, cost overruns, and even giving the technological advantage to our competitors.

When we introduce a product, we share the technology with the marketplace. If we implement poorly, we damage our reputation, while giving the competition the chance to copy the technology we have spent so much to develop. Through poor implementation, we also tell our competition what *not* to do when they copy us.

Technology can be optimized using robust design techniques, just like products and processes. The difference is that we are seeking to open the operating window in a generic context, *independent* of a specific product.

When we optimize a technology for a particular product, it is so tightly coupled with the product that it is difficult to use the same technology with another similar product. As a result, much tuning may be required to make the second product work with the new, "proven" technology.

It may seem easier to develop technology in a specific application because we have known performance targets and a more concrete objective. Our challenge is to develop technology without knowing the desired target or the exact application.

It is more effective (and faster) to develop technology off-line, independent of a specific product. Such technology development may be likened to "stocking our bookshelf with technological capabilities." We need a wide range of options to give us the flexibility to apply the best technology for the application.

The focus of the technology development is to open the operating window and identify tuning factors that can be adjusted to a specific product application. We will ultimately use those tuning factors to help us implement the technology on a specific application, at the desired performance levels.

Tuning factors are control factors that are chosen to easily adjust the technology to a specific application. While any control factor could be used, some are more practical (easier or less costly) than others. This phase of technology development is called tuning research.

The steps of technology development are:

- Define the ideal function.
- Determine control and noise factors.
- Optimize by maximizing the S/N Ratio using parameter design.
- Choose appropriate tuning factors.

The overall approach focuses on discovering, as much as possible, how the technology behaves. Physical testing is often used to begin creation of mathematical models.

By taking the time to understand the technology being developed, there is a far greater opportunity to implement successfully. This is especially important if we are dealing with the technologies that form the core competence of our corporation—one of the key components of our competitive leadership.

We can even increase our technological lead because our competition will be trying to copy our technology without our level of understanding of it. Competitors may copy our technology and introduce some of the problems we avoided.

DESIGNING THE RIGHT PRODUCT

To this point, we have discussed the development process, assuming that we already have a well-defined product idea in mind. While most organizations begin product development in this way, many do not have an adequate understanding of the customer wants, needs, and expectations to define a product adequately. Without thorough knowledge of the customer, good products can be designed—they just may not sell.

Our engineers have their own paradigms, or rules of operation, that influence their approach to designing a product. These paradigms drive their perception of good and bad features, functions, and performance.

The customer has a different set of paradigms that influences his or her perception of good and bad features, functions, and performance. Which one is right?

FIGURE 10–13

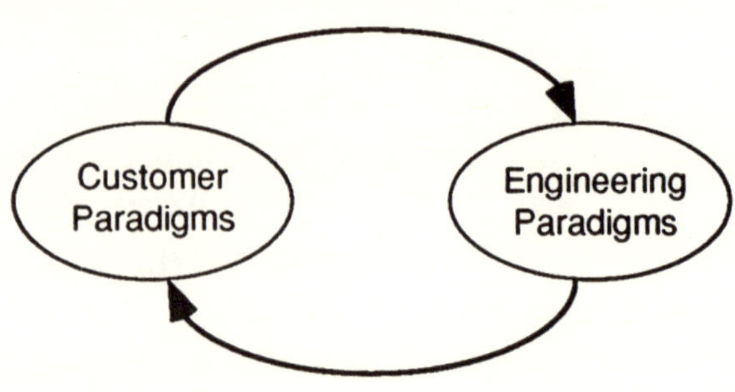

Perhaps both, but the one that really matters is the customer's paradigm, because the customer will decide the market success of a product.

If our engineering paradigms dominate and prevent us from satisfying the customer's paradigms, we design our products with a technological bias, missing the customer wants.

In general, we will want to adapt our engineering paradigms to comprehend the customer paradigms (see Figure 10–13). In this manner, we greatly increase the likelihood that the function we define and design will meet the customer intent.

Otherwise, we are obliged to try to "convince" the customer (through marketing or incentives) that our product is really what he or she wants. The greater the difference between the customer paradigm and the engineering paradigm, the harder this will be. It is very difficult to "educate customers" to change their paradigms. Most customers will change brands rather than change paradigms.

Occasionally, we have the opportunity to introduce an innovative product that requires a shift in the cus-

FIGURE 10–14

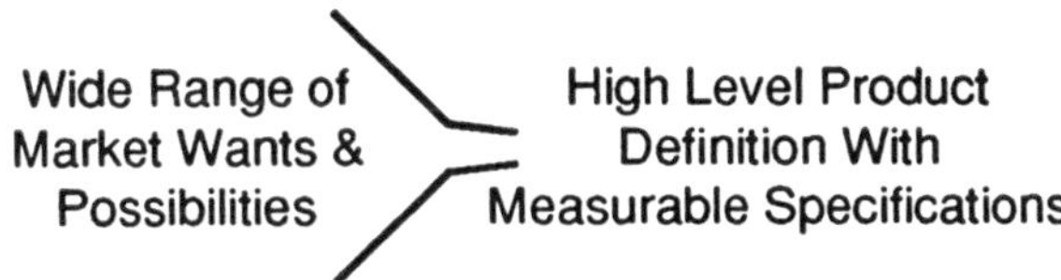

tomer's paradigm to create our market. When disposable diapers were first introduced, the customer paradigm was "if you love your baby, you'll use cloth diapers." It took years and millions of dollars of marketing to overcome that paradigm. If we must change the customer paradigm, it is easier to do so based on a thorough knowledge of the customer wants.

It is imperative that our product decisions be made based on a firm understanding of the voice of the customer. While we cannot always give customers all that they want, if we better satisfy customer wants in a commercially viable manner, we will have a competitive advantage.

QFD is an approach to help us to understand the meaning behind the "voice of the customer" and translate that meaning into overall product specifications, as shown in Figure 10-14.

We can think of QFD as helping us to consider the many product possibilities and create an engineering specification for products that will meet our customer wants in a commercially viable manner.

QFD uses matrix charts (see Figure 10–15) to help translate the voice of the customer (want #1 through #4), into engineering specifications (spec. #1 through

FIGURE 10–15

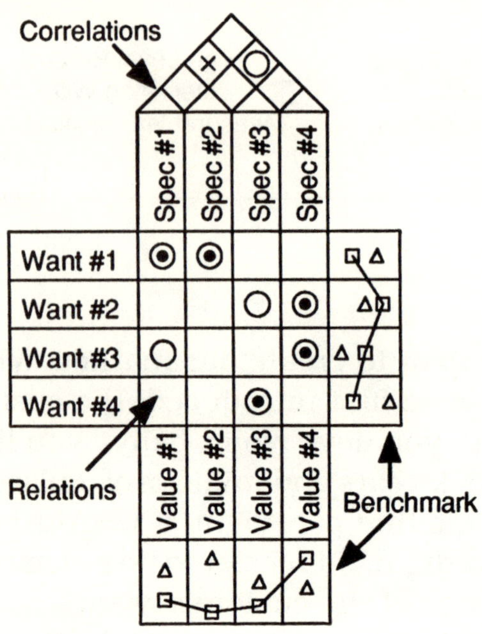

#4) and their values (value #1 through #4) at the overall product level.

The customer wants are often vague and ambiguous. Our challenge is to understand them and convert them into clear and measurable specifications that are understood throughout our organization.

The relations describe the cause-effect relationship between specifications and customer wants. Double circles depict strong relations, while circles depict weaker relations. In this case, if we achieve spec. #3 at value #3, we expect to achieve the customer want #4.

The benchmark information graphically describes the competitive reality of today's marketplace so that we may make informed decisions about how to position our new product. It also helps us to check the correctness of our proposed specifications.

The correlations describe how the specifications work together. If they reinforce each other (specs. #2 and #4), we draw a circle. If they conflict with each other (specs. #1 and #3), we draw a cross. Some of these conflicts can drive our research efforts to seek a technological solution for a future product.

QFD uses other matrix charts to translate product specifications into part specifications, process parameter values, and shop floor control plans.

PUTTING THE TOOLS TOGETHER

We can think of QFD as helping us to focus on doing the right things, while robust design helps us to do things right. Clearly, both are necessary for commercial success.

Returning to our bottle model (see Figure 10–16), we see that the voice of the customer is taken as input to determine a focused set of product specifications. The best design concept is chosen to satisfy these specifications at the lowest cost. The design concept is then developed to maximize design latitudes through robust design practices. Production processes are concurrently optimized to minimize product to product variation. The net design latitude is maximized to allow latitude in actual product operation (customer use, environments, degradation, etc.).

While the approach typically involves some new tools, many of the tools we use today continue to play an important role.

FIGURE 10–16

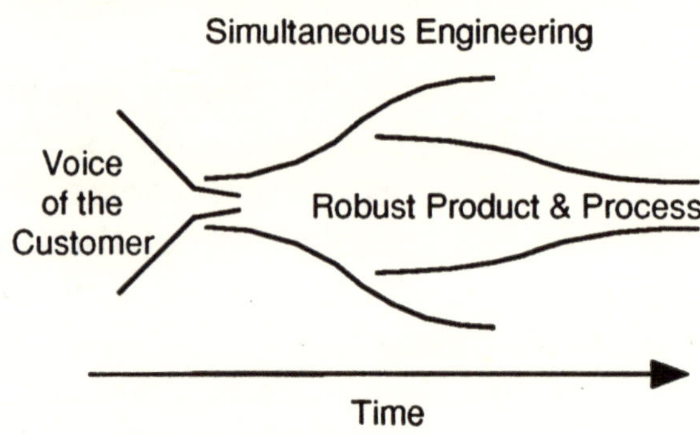

Figure 10–17 illustrates how some of these tools work together to promote rapid product development.

Through our strategic planning efforts, we choose our overall company direction, determine the customers and markets we wish to serve, and define the core competencies we will exploit.

Through phase one of QFD (also called the House of Quality), we seek to understand the customers and define the overall specifications for specific products.

Conflicts between specifications may drive R&D efforts to find a technological solution to the conflict. Note that the technology development happens off-line—not as a direct part of the specific product development. It serves the entire business, not just a specific product.

Having defined the product specifications, we then seek to create the best product concept—one that will best satisfy the specifications (and hence the customer wants) at the lowest possible cost.

These ideas come from our internal creativity, competitive benchmarking, DFA, VE, and our technology

FIGURE 10–17

Strategic Planning
Key Corporate Technologies
Product Technology Development
Core Robust Technologies
Existing Technology
Competitive Benchmarking
DFA
VE
Product Technology Bookshelf
Cost & Other Business Issues
Who is the Customer?
Market Segments
Voice of the Customer
QFD-1
Product Spec's
Customer Wants
Spec Values
Competitive Benchmarking
Selection Criteria
Concept Alternatives
Scores
Pugh Concept Selection
DFA
VE
Function Tree & Bill of Materials
QFD-2
Part / Mat'l Spec's
Product Spec's
Spec Values
Parameter Design & Tolerance Design
Environments, Use Conditions, Durability Expectations
Noise Factors
Ideal Functions
Continue Similar Approach for Process & Production Planning

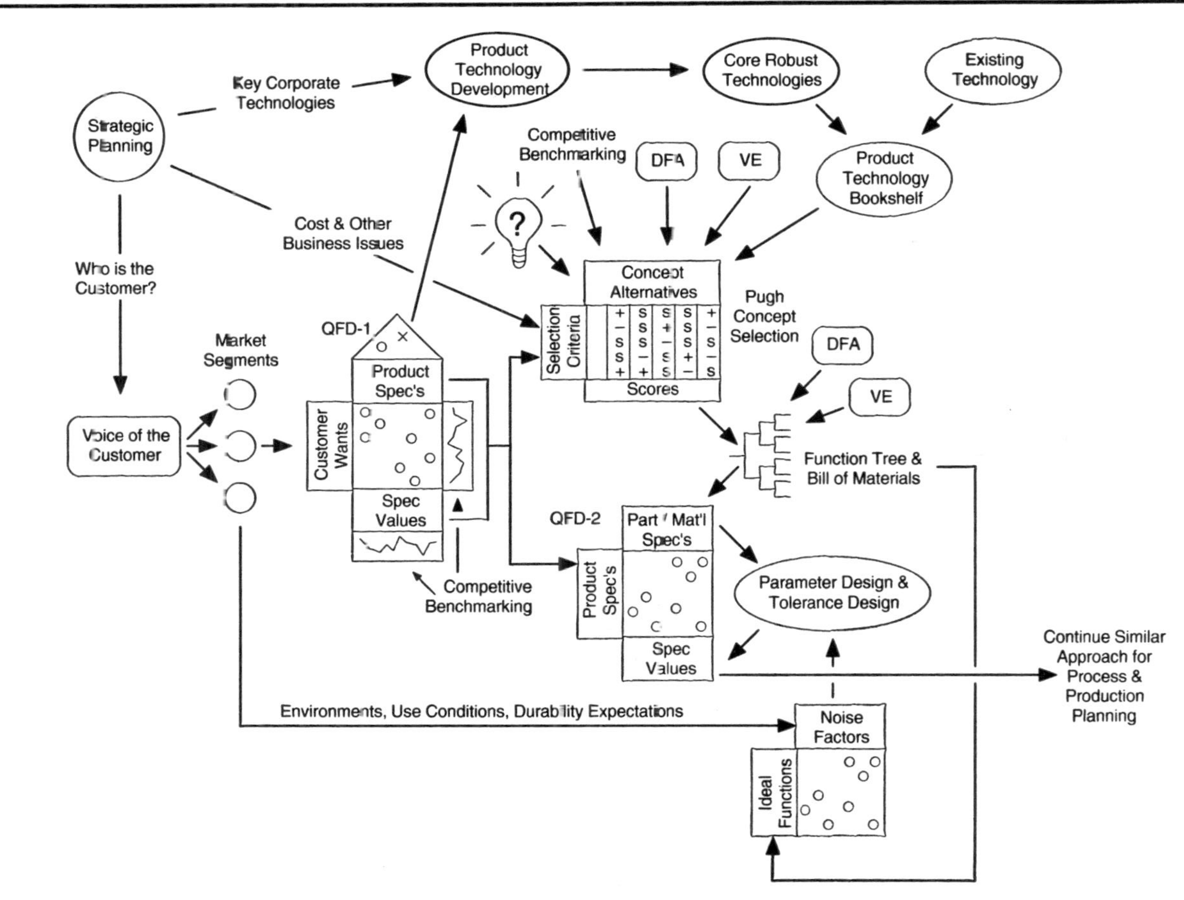

"bookshelf." Through Pugh Concept Selection, we seek to synthesize a single concept that is better than any of our other ideas.

This concept is then detailed into a bill of materials, using function trees, DFA, and VE to maximize function in the simplest, lowest-cost manner.

The product specifications are carried from QFD phase one to phase two of the QFD process. The critical part and material specification from the bill of material are determined and fed into phase two.

Through Taguchi's parameter design and tolerance design approaches, we determine the best value and tolerances for each of the critical control factors in the design. Input to the parameter design includes noise factors identified by earlier customer research.

Process development is handled in a similar manner. Our choices of process alternatives may be constrained because we often must use existing equipment. Noise factors will typically include factory, environmental, variable, and lot-to-lot material variation.

It is vitally important to remember that in a simultaneous engineering environment, these steps (especially product and process design) happen concurrently. They feedforward and back, but must happen together.

CHANGING OUR PARADIGMS

Using the tools is not enough—*how* we use them is vitally important. A hammer can be used to drive nails or put holes in walls. The method makes the difference.

To reduce product development cycle time, there is a fundamental paradigm shift required.

This shift is from minimizing problems to maximizing function. While both approaches will work, given enough time, we cannot afford the time needed to find and fix all the problems.

The implications of maximizing function are:

- Understanding customer wants.
- Translating wants to specifications.
- Defining the ideal function.
  At lowest level of detail.
  Energy transfer.
- Choosing the best design concept.
  Developed technology.
  Simplify design.
- Product testing for knowledge gain.
  Measurable results.
  For optimizing product.
  For building math models.
  Not for finding problems.
- Optimizing design for robustness.
  Identify control factors.
  Identify noise factors.
  Choose control factor values.
  Maximize S/N Ratio.
- Prototype testing for confirmation.
  Not for debugging.
  Represents production intent.
  Design work is done.
  Single phase (this is it!).
- Optimizing process for robustness.
  Identify control factors.
  Identify noise factors.
  Choose control factor values.
  Maximize S/N Ratio.
- Developing technology off-line.

FIGURE 10-18

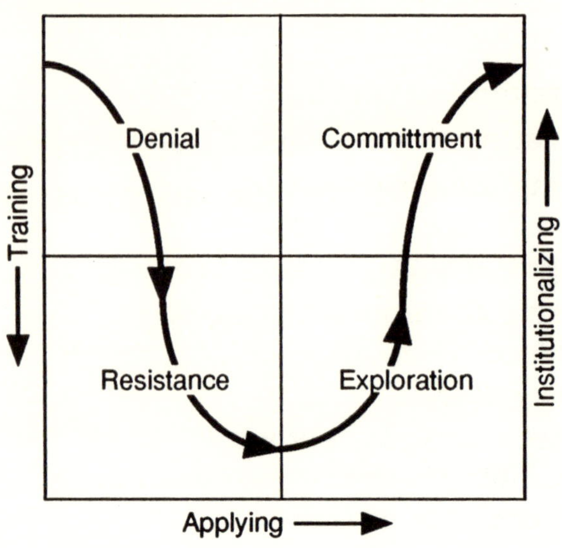

Identify ideal function.
Identify control factors.
Identify noise factors.
Choose tuning factors.
Build math model.
Maximize S/N Ratio.

Even if we accept all of these implications, the paradigm shift will require that we deal with the "people issues." Cynthia Scott and Dennis Jaffe of Flora/Elkind Associates discovered that people tend to go through the stages shown in Figure 10–18 when faced with change.

Many who fail to see the need for change deny it entirely. They can't believe it's real. It's probably just another "program of the month."

After seeing that the change will not go away, individuals resist, often employing the following "justifications": "What's wrong with the way we've been doing it?" "We're already pretty successful." "This will just make more work!" "I'm already too busy!"

After seeing some small merit in the change, there is a willingness to explore it. "Let's try it out!" becomes a common expression.

After experiencing the value of the change, people become committed to it and help others to make the change.

Different people go through this transition at different rates, but they do not go from denial to commitment without the intervening steps. When senior management orders a change and the organization falls all over itself, it's a good bet that there is a lot of hidden denial and resistance. We shouldn't confuse following orders with commitment.

A more successful approach is to lead the organization through the change process and make successes highly visible, letting the visible facts overcome resistance. Training is often necessary to help the transition from denial to resistance. Resistance gives way to exploration as we apply the new approach and see that it may have some merit. Institutionalization occurs as many of the explorations become visible successes and a "critical mass" of people becomes fully committed to the new approach.

Because some people go through the transition faster, it is usually easier to start with those who are more open to change.

Figure 10–19 is a common profile of the willingness to accept change in organizations. The innovators will try almost anything if it's new. The risk takers will try

FIGURE 10–19

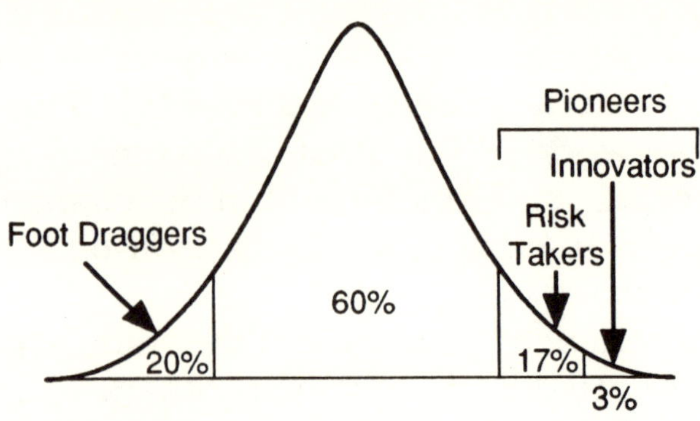

it if it seems reasonable. These two groups are the "pioneers" who will most readily change paradigms. The foot draggers are the guardians of the status quo. Everyone else is in the middle.

We usually know who the pioneers are. If we identify these receptive individuals and recruit them to help make the initial change, we can progress faster. Even the pioneers will generally need some training. Enthusiasm must be coupled with knowledge to be successful.

The knowledge required to adopt the new engineering paradigm will include:

- Systems thinking and design practice.
- Analytical methods for rapid progress.
- Measurement approaches to quickly get useful engineering information.
- Decision-making approaches.
- Variation and how to deal with it.
- Parameter design application.

Pioneering managers will also need to understand the behavioral aspect of the change process.

SO WHAT DO I DO TOMORROW?

To maximize our ability to implement rapid product development, we must significantly change the way we manage our product development. This requires changes at the highest management levels to be completely effective. Nevertheless, there are positive actions that individual engineers can take to achieve robust design on a smaller scale.

Individual Actions for Pioneers

- Learn all you can about how the customer sees your product.
- Learn all you can about your competitors (design, manufacturing, etc.).
- Learn how customers perceive the difference between your product and that of your competition.
- As you develop your products, seek to determine the ideal function.
- Use robust design approaches to deal with noise (avoid compensation if at all possible).
- Seek to revise attribute tests to obtain measurable results.
- Create math models for your products.
- Use VE and DFA approaches to create the simplest design to maximize the ideal function.
- When problems are found, look for energy-related causes.

Actions for Pioneering Senior Management

- Restructure development for a single-phase prototype.
- Instill the discipline to make the prototypes truly production intent.
- Separate technology development from product development; provide separate funding.
- Provide avenues for engineers to receive *first-hand* input of customer wants.
- Review the process by which products are developed; ask for visible proof that robust design practices are being used.

Implementation Realities

Even a group of enthusiastic, knowledgeable pioneers will often run directly into the immune system of the organization. This immune system helps preserve the status quo by resisting unusual changes. This can be useful to maintain stability, but it's harmful when trying to make a deliberate change!

A pioneering senior manager will often be needed to provide support and break down barriers faced by the pioneers. This manager also fosters an environment in which risk taking is favored, with minor failures being simply the price of progress. This is especially important, for if management will not tolerate failures, even the pioneers will quickly "wise up" and avoid risks.

As successes are achieved, they should be made highly visible. This will help to:

- Provide recognition to the pioneers.
- Educate the organization.
- Convince the doubters.
- Build additional support.

The continued change will simply not be tolerated if results are not apparent. By making the successes visible, we help maintain the momentum. If we don't, interest and support will dwindle.

Shifting our engineering paradigms is a difficult prospect, requiring substantial investment in the human side of our business. With proper care and nurture, the return on this investment will exceed anything Wall Street has to offer.

ACKNOWLEDGMENTS

Dr Donald P. Clausing, of MIT, one of the American pioneers of Dr Taguchi's approaches, for originating product-development concepts and the bottle model.

Dr Stuart Pugh, of the University of Strathclyde, for formalizing some of the approaches in conceptual design selection and development (Pugh Concept Selection).

Cynthia Scott and Dennis Jaffe, of Flora/Elkind Associates, for creating the four-quadrant change model.

Rick Pedi, of Pedi, Moesta & Associates, for contributing the strategic planning thinking to the graphic on page 199.

Dr Genichi Taguchi for developing the parameter design approach to robustness, rapid technology development, and the body of knowledge we call Taguchi Methods™.

Keith B. Termaat, of Ford Motor Company, for conceiving the "funnel" preceding the bottle, depicting the narrowing of market possibilities into specifications that define a particular product.

James Wilkins and Allen Chartier, of ASI, for initially editing this paper.

Special thanks to members of the Ford Design Institute Advisor Panel for the review and critique that made this paper possible.

REFERENCES

Bebb, B. "Quality Engineering Design as Major Determinant of Time-to-Market and Product Cycle Time," *Designing the Future,* joint workshop, Xerox Corp. and Design Productivity Center, February 1988.

Clausing, D. "Concurrent Engineering," Design and Productivity International Conference, February 1991.

Clausing, D., and S. Pugh. "Enhanced Quality Function Deployment," Design and Productivity International Conference, February 1991.

Clausing, D. "Flexible Product Development," MIT paper, May 17, 1991.

Eureka, W. E., and N. E. Ryan. *The Customer-Driven Company: Managerial Perspectives on QFD,* Irwin Professional Publishing and the ASI Press, 1994.

Hauser, J., and D. Clausing. "The House of Quality," *Harvard Business Review,* May–June 1988.

Kuo, W., and J. Hsu. "Update: Simultaneous Engineering Design in Japan," *IE,* October 1990.

Mori, T. *The New Experimental Design,* ASI Press, Allen Park, Michigan, 1990.

Phadke, M. *Quality Engineering Using Robot Design*, Prentice Hall, Englewood Cliffs, N.J., 1989.

Prahalad, C., and G. Hamel. "The Core Competence of the Corporation," *Harvard Business Review*, May–June, 1990.

Scott, C., and D. Jaffe. *Mastering Change*, Elkind/Flora Associates, 1991.

Stickler, M., and E. Turcotte. "Quality Begins at the Beginning: Product and Process Design," *AIAG Actionline*, August 1991.

Sullivan, L. "Quality Function Deployment," *Quality Progress*, June 1986.

Taguchi, G. *Introduction to Quality Engineering*, Asian Productivity Organization, 1986, and *System of Experimental Design: Engineering Methods to Optimize Quality and Minimize Cost*, UNIPUB/Kraus Int'l., 1987.

Glossary

Company measures: Global measures not specified to design that translate the voice of the customer into broad, actionable issues the company can work on in response to customer wants. Also called technical requirements.

Companywide Quality Control: See Total Quality Management.

Competitive assessment: A benchmarking process for evaluating customer requirements and company measures that determines the performance of the competition.

Concurrent engineering: See simultaneous engineering. (The term simultaneous engineering is favored by the automotive industry, concurrent engineering by the defense industry.)

Correlation: The co-relationship between any two company measures. It represents a team judgment about whether action to improve one company measure will harm or help the other company measure.

Customer rating: Customer opinion of the importance of each customer want, typically rated on a numerical scale.

Customer requirements: Written representations of the voice of the customer: wants and needs of customers determined during interviews and through surveys, and so on.

Deployment: The process of transforming information through the product development process into a salable product.

Design for Assembly (DFA): A method of design simplification that focuses on reducing assembly complexity without sacrificing product performance. DFA was introduced by Geoffrey Boothroyd and Peter Dewhurst of the University of Massachusetts in 1980.

Design of Experiments (DOE): Statistical techniques used to discover the effects of factors upon desired results. Taguchi Methods use a simple subset of DOE to get engineering results quickly and without extensive statistical expertise.

Failure-Mode and Effect Analysis (FMEA): A method for failure reduction that identifies potential product or process weakness, assesses risks, and determines preventive actions.

Fault Tree Analysis (FTA): A method of tracing failures through a chain of cause-and-effect relationships to their root cause. FTA can provide a means to estimate failure probability.

House of Quality: A term used to describe the QFD product-planning matrix designed to increase customer satisfaction to improve quality. The correlation matrix at the top gives the appearance of a house. Its purpose and appearance together account for the term.

Matrix: A rectangular chart where two sets of items are compared. One set is entered horizontally and the second set vertically.

Policy Management: A process for developing achievable business plans and then deploying them throughout the company that closely resembles QFD. Also called Policy Deployment.

Pugh Concept Selection: A creative method for comparing design concepts that seeks to synthesize a single concept better than any of the initial concepts. Devised by Dr Stuart Pugh, a British engineer.

Quality Function Deployment: A system for translating customer requirements into appropriate customer requirements at each stage of the product-development cycle, from research and development to engineering, manufacturing, marketing, sales, and distribution.

Relationships: Symbolic representations of cause-effect phenomenon.

Reverse Fault Tree Analysis (RFTA): A "mirror image" of Fault Tree Analysis (FTA) that focuses on drivers for success rather than causes of failure. An FTA can be logically transformed into an RFTA.

Robust design: The use of Taguchi Methods to define and optimize product function.

Seven Basic Tools: A set of seven tools primarily used to derive information from quantitative data. The tools are cause-and-effect diagram, Pareto chart, check sheet, scatter diagram, histogram, flow chart, and control chart.

Seven Management Tools: A set of seven tools primarily used to derive information from qualitative data. The tools are affinity diagram, relationship digraph, tree diagram, matrix chart, matrix data analysis, process design program chart, and arrow diagram.

Simultaneous engineering: A way of simultaneously designing products and the processes for manufacturing those products through the use of parallel efforts to ensure manufacturability and reduce cycle time. Also called concurrent engineering.

Taguchi Methods™: An engineering approach that optimizes the product to achieve stable, cost-effective performance.

Target value: A goal established for each company measure. It represents the team judgment of the level of achievement that must be attained for the company measure to satisfy customers.

Technical requirements: See company measures.

Total Quality Management (TQM): A management approach that integrates fundamental management techniques, existing improvement efforts, and technical tools to transform a company into a continuously improving organization. Also called Companywide Quality Control and Total Quality Control.

Value Analysis/Value Engineering (VAVE): A method of design simplification that focuses on reducing cost while maintaining or improving product functionality. This was introduced by Larry Miles of General Electric Corp. in 1947.

Voice of the customer: The wants and needs that customers express during interviews and surveys relative to products and service performance.

Bibliography

Akao, Y., ed. *Quality Function Deployment: Integrating Customer Requirements into Product Design.* Cambridge, MA: Productivity Press, 1990.

Bossert, J. L. *Quality Function Deployment: A Practitioner's Approach.* Milwaukee: ASQC Quality Press, 1990.

Camp, R. C. *Benchmarking: The Search for Industry Best Practices That Lead to Superior Performance.* White Plains, NY: Quality Resources, 1989.

Clausing, D. P., and S. Pugh. "Enhanced Quality Function Deployment." Paper presented at Design and Productivity International Conference, 1991.

Cohen, L. "Quality Function Deployment: An Application Perspective from Digital Equipment Corporation." *National Productivity Review*, Summer 1988.

Covey, S. R. *The Seven Habits of Highly Effective People: Powerful Lessons in Personal Change.* New York: Simon & Schuster, 1989.

Ealey, L. A. *Quality by Design: Taguchi Methods and U.S. Industry,* 2nd ed. Dearborn, MI: ASI Press; and Burr Ridge, IL: Irwin Professional Publishing, 1994.

Eureka, W. E., and N. E. Ryan. *The Customer-Driven Company: Managerial Perspectives on QFD,* 2nd ed. Dearborn, MI: ASI Press; and Burr Ridge, IL: Irwin Professional Publishing, 1994.

Hauser, J. R., and D. Clausing. "The House of Quality." *Harvard Business Review*, May–June, 1988.

Introduction to Quality Engineering Implementation Manual. Dearborn, MI: ASI Press, 1992.

King, B. *Better Designs in Half the Time: Implementing QFD in America.* Methuen, MA: GOAL/QPC, 1987.

LeBouef, M. *How to Win Customers and Keep Them for Life.* New York: Berkeley Publishing Group, 1989.

Marsh, S.; J. W. Moran; S. Nakul; and G. Hoffherr. *Facilitating and Training in Quality Function Deployment.* Methuen, MA: GOAL/QPC, 1991.

Mori, T. *The New Experimental Design: Taguchi's Approach to Quality Engineering.* Dearborn, MI: ASI Press, 1990.

Orthogonal Arrays and Linear Graphs. Dearborn, MI: ASI Press, 1986.

Phadke, M. *Quality Engineering and Robust Design.* New York: Prentice-Hall, 1989.

Proceedings, First through /Tenth Symposiums on Taguchi Methods. Dearborn, MI: ASI Press, 1984–93.

Pugh, S. Concept Selection—The Method That Works. In Proceedings of ICED, 1981.

————. *Total Design: Integrated Methods for Successful Product Engineering.* Reading, MA: Addison-Wesley, 1990.

Quality Engineering Executive Briefing. Dearborn, MI: ASI Press, 1988.

Quality Function Deployment and the Competitive Challenge: A Special Nine-Part Video Series with Bill Eureka. Dearborn, MI: ASI Press, 1989.

Quality Function Deployment: The Budd Co. Case Study. Dearborn, MI: ASI Press, 1987. Videotape.

Quality Function Deployment Executive Briefing. Dearborn, MI: ASI Press, 1992.

Quality Function Deployment for Continuous Batch Process Implementation Manual. Dearborn, MI: ASI Press, 1992.

Quality Function Deployment for Hardware Implementation Manual. Dearborn, MI: ASI Press, 1992.

Quality Function Deployment for Service Implementation Manual. Dearborn, MI: ASI Press, 1992.

Quality Function Deployment: The Kelsey-Hayes Case Study. Dearborn, MI: ASI Press, 1987. Videotape.

Rosenthal, S. R. *Effective Product Design and Development: How to Cut Lead Time and Increase Customer Satisfaction.* Burr Ridge, IL: Irwin Professional Publishing, 1992.

Scholtes, P. R. *The Team Handbook: How to Use Teams to Improve Quality.* Madison, WI: Joiner Associates, 1988.

Shunk, D. L. *Integrated Process Design and Development.* Burr Ridge, IL: Irwin Professional Publishing, 1992.

Sudman, S., and N. E. Bradburn. *Asking Questions: A Practical Guide to Questionnaire Design.* San Francisco: Jossey-Bass Inc., 1982.

Sullivan, L. P. Quality Function Deployment. *Quality Progress,* June, 1986.

Taguchi, G. *Introduction to Quality Engineering.* Tokyo: Asian Productivity Organization, 1986.

———. *System of Experimental Design: Engineering Methods to Optimize Quality and Minimize Cost.* White Plains, NY: UNIPUB/Kraus Int'l, 1987.

———. *Taguchi Methods: Case Studies from Japan.* Dearborn, MI: ASI Press, 1993.

———. *Taguchi Methods: Case Studies from U.S. and Europe.* Dearborn, MI: ASI Press, 1993.

———. *Taguchi Methods: Case Studies in Measurement from Japan.* Dearborn, MI: ASI Press, 1993.

———. *Taguchi Methods: On-line Production.* Dearborn, MI: ASI Press, 1993.

———. *Taguchi Methods: Research and Development.* Dearborn, MI: ASI Press, 1993.

———. *Taguchi Methods: Signal-to-Noise Ratio for Quality Evaluation.* Dearborn, MI: ASI Press, 1993.

———. *Taguchi Methods: Case Studies from Japan.* Dearborn, MI: ASI Press, 1993.

Taguchi Methods: Selected Papers on Methodology and Applications. Dearborn, MI: ASI Press, 1988.

ter Haar, S.; D. P. Clausing; and S. D. Eppinger. *Integration of Quality Function Deployment and the Design Structure Matrix.* Massachusetts Institute of Technology, 1993.

Transations. First through Fifth Symposiums on Quality Function Deployment. Dearborn, MI: ASI Press, and Methuen, MA: GOAL/QPC, 1989–93.

Wheelwright, S. C., and K. B. Clark. *Revolutionizing Product Development, Quantum Leaps in Speed, Efficiency, and Quality.* New York: The Free Press, 1992.

Ziaja, H. J. "Total Product Quality Model." Paper presented at ASQC 44th Annual Quality Congress, 1990.

Thank you for choosing Irwin Professional Publishing (formerly Business One Irwin) for your information needs. If you are part of a corporation, professional association, or government agency, consider our newest option: Custom Publishing. This service helps you create customized books, manuals, and other materials from your organization's resources, select chapters of our books, or both.

Irwin Professional Publishing books are also excellent resources for training/ educational programs, premiums, and incentives. For information on volume discounts or Custom Publishing, call 1-800-634-3966.

Other books of interest to you from Irwin Professional Publishing . . .

THE TQM ALMANAC
1994–95 Edition
Timeplace, Inc.

An all-encompassing guide to quality-related materials, resources, and information that quality managers need to start and maintain successful TQM programs. This time-saving guide to TQM books, videos, software, consultants, seminars, conferences, and more also provides an overview of the current trends in the quality improvement field.
ISBN: 0-7863-0242-9

THE ISO 9000 ALMANAC
1994–95 Edition
Timeplace, Inc.

Discover all the resources you need for ISO 9000 registration! Filled with time-saving information, you'll find a comprehensive listing of consultants, videos, seminars, books, articles, and much more to ensure registration success!
ISBN: 0-7863-0243-7

SYNCHROSERVICE!
The Innovative Way to Build a Dynasty of Customers
Richard J. Schonberger and Edward M. Knod, Jr.

From the best-selling author of *Building a Chain of Customers*! Schonberger and Knod give you their latest ground-breaking strategy—synchroservice—to help your company ensure an organization-wide commitment to seamless, consistent, customer-driven service for enhanced customer loyalty.
ISBN: 0-7863-0245-3

WHY TQM FAILS AND WHAT TO DO ABOUT IT
Mark Graham Brown, Darcy E. Hitchcock, and Marsha L. Willard
Co-published with the Association for Quality and Participation

Finally—the answers to solve the troubles surrounding TQM! This trouble-shooting guide shows how to avoid the common mistakes made by organizations during each crucial phase of the TQM implementation process. Whether your organization is in the midst of a TQM crisis or just wants to avoid disappointing results, this pioneering reference offers practical, concrete guidance to make TQM succeed at your company!
ISBN: 0-7863-0140-6

GLOBAL QUALITY
A Synthesis of the World's Best Management Methods
Richard Tabor Greene
Co-published with ASQC Quality Press

This comprehensive resource organizes the chaos of quality improvement techniques so you can identify the best approaches for your organization. Includes the 24 quality approaches used worldwide, the essentials of process reengineering, software techniques, and seven new quality improvement techniques being tested in Japan.
ISBN: 1-55623-915-7

Available at bookstores and libraries everywhere.

Please respond so we can continue to serve you!

If you enjoyed this book, continue your learning through other ASI sponsored educational opportunities. Please send me more information on:

- [ ] ASI Publications — More than 100 textbook, courseware, and software items.
- [ ] ASI Training Courses — Industry leading techniques for the improvement of products and services.
- [ ] Complete ASI Information Package — Including publications, course offerings, in-house programs, and more.
- [ ] I am interested specifically in: _______________________________________

Name: _______________________________________

Title: _______________________________________

Company: _______________________________________

Address: _______________________________________

City: _________________________ State: _____________ Zip: _____________

Phone: _______________________________________

AMERICAN SUPPLIER INSTITUTE
(313) 336-8877 • 1-800-462-4500